Calendar Problems

from the

Mathematics Teacher

John Grant McLoughlin
University of New Brunswick
Fredericton, New Brunswick

NCTM®

NATIONAL COUNCIL OF
TEACHERS OF MATHEMATICS

Copyright © 2002 by

THE NATIONAL COUNCIL OF TEACHERS OF MATHEMATICS, INC.

1906 Association Drive, Reston, VA 20191-1502
(703) 620-9840; (800) 235-7566; www.nctm.org

Second printing 2006

ISBN: 0-87353-515-4

The National Council of Teachers of Mathematics is a public voice of mathematics education, providing vision, leadership, and professional development to support teachers in ensuring mathematics learning of the highest quality for all students.

Printed in the United States of America

Contents

Preface

For many years, the National Council of Teachers of Mathematics (NCTM) has published the *Mathematics Teacher*, a journal intended mainly for a secondary-level audience. A prominent feature of the journal is a monthly calendar of problems. Each calendar presents a problem for students and readers to solve each day. All problems submitted for the calendars are reviewed by three people. As the editor of the problems for two years, I was expected to gather the reviewed contributions and prepare calendars, along with detailed solutions, for the months from October through May (the September calendars typically feature problems selected from mathematics contests). In addition, as editor, I handled a considerable amount of mail offering feedback or alternate methods of solution. Letters suitable for publication appear in Readers' Reflections in subsequent issues. Detailed solutions accompany each calendar.

Several years ago, Harry Tunis and Joan Armistead approached me with a proposal to prepare a book that could serve as a resource for teachers. The idea for such a book struck me as a good one, and I began considering five years' worth of problems—about 1200 problems in total, published between October 1993 and May 1998. I selected nearly 400 of these to include in the book. To organize and arrange the problems, I imposed on them a classification structure that would allow teachers to identify problems readily and clearly in the areas of number theory, coordinate geometry, spatial sense, and so on. In addition, I prepared solutions, approximately half of which are notably different from those that appeared in the original published calendars.

The calendar is a popular feature of the *Mathematics Teacher*. The book is intended to present this feature in a new format—as a collection of problems organized by topic. The hope is that the book may reach a new and larger audience while serving as a valuable tool for promoting problem solving and mathematical entertainment among students, educators, and others interested in mathematics.

Many people have contributed in different ways to this book. Its preparation was generously supported by the efforts of students and staff at Memorial University of Newfoundland. The author wishes to extend appreciation to the many students who have assisted at various stages of the book's development: Natasha Blanchard, Courtney Clarke, Melissa Langille, Tim Manning, Renée Nolan, and Bronwyn Rideout. Boyde Hann's contribution is particularly noteworthy, since he assisted in the selection of the problems as well as the overall organization of the manuscript in its initial stages. Deanne Burton and Renée Lynch have also been longstanding contributors to the project. Both have generously devoted many hours to proofreading, checking sources, referencing, and attending to other necessary tasks. The office staff, consisting of Carolyn Bourne, Wanda

Kitchen, Carolyn Lono, Eileen Ryan, and Laura Walsh, has been very helpful at numerous stages of the process. Jackie Pitcher-March has generously made arrangements for student assistants to contribute to the project.

I am also grateful to the staff at the National Council of Teachers of Mathematics. Harry Tunis has been my primary link to NCTM throughout this project. Joan Armistead has acted as my editorial contact and advisor while I have served as a calendar reviewer, editor, and author over the past thirteen years. Anita Draper edited the copy of the book, and Jo Handerson designed it and laid out the pages, with assistance from Jodi Flood, who worked on illustrations for a number of the problems. In addition, I would like to thank Sue Brown and Terry Coes, the *Mathematics Teacher* editorial panel members with whom I worked as an editor. Also, the preceding editors, Margaret Raub-Hunt and William Hunt, assisted me in making the transition to editor. Moreover, the subsequent two editors, Monte Zerger and Scott Stull, served as exemplary reviewers during my term as editor.

I cannot close any expression of gratitude without thanking Kathy, Moira, and Heather for their patience and support.

Finally, I would like to add my assurance that a conscientious effort has been made to see that proper credit has been given to the contributors and sources of the problems that appear in the book. Reasonable steps have also been taken to minimize any errors that might diminish the value of the plethora of problems, solutions, and contributions presented here. Feedback from readers of the book is most welcome. No problem should be seen as closed. Alternate solutions or commentary on the problems would be appreciated. Enjoy the problem solving!

John Grant McLoughlin
Faculty of Education
University of New Brunswick
Fredericton, NB E3B 5A3
johngm@unb.ca

Introduction

A brief overview of the layout of the book may assist readers in using it to best advantage. Although the core of the book consists of nearly four hundred mathematical problems, two pieces that have been placed ahead of the problems may increase the readers' understanding of the context in which the book appears.

Seeing a sample calendar may help put the book into perspective. The January 1995 calendar, which was prepared by Richard Brown, has been included here in its entirety to give readers an idea of how all the problems looked in their original calendar setting in the *Mathematics Teacher*. This specimen calendar may also give readers some appreciation for the range of topics that a single calendar may embrace. The January 1995 calendar featured a particularly rich mixture of problems that work better together than as separate entities. The calendar made an immediate impression on me as one of its reviewers at the time.

A section called Readers' Favorites has been placed after the January 1995 calendar and before the problems in the book. This section focuses on five problems that stimulated readers to write and share an assortment of insights and solutions. Any of these five problems may prove to be ideal for promoting discussion in a variety of mathematical contexts. The problems and the accompanying comments depict the potential for mathematical development that a good problem may enfold and that tackling such a problem may realize. The selection highlights these problems rather than hiding them amidst the collection. Nevertheless, many such gems await discovery in the collection! The Readers' Favorites section need not be read before approaching the problems but can be read at any time.

The problems themselves have been separated into fourteen sections corresponding to different mathematical topics. Readers are welcome to take any of the fourteen problem sections as an entry point and dive into the book! Specific mathematical areas are represented in the first ten sections: Number Theory, Coordinate Geometry, Spatial Sense, Logic, Algebra, Probability, Geometry, Logs and Exponents, Sequences and Patterns, and Counting. The final four sections—Word Problems, Puzzles and Games, Facts, and Quickies—are less easily defined with respect to the mathematical content of the problems assigned to them. The first two of these titles require minimal explanation, and the problems that are included in the Facts section are generally connected to some aspect of the history of mathematics. However, the Quickies section, which contains more than 20 percent of the problems in the book, is unique, offering a miscellany of problems from various areas of mathematics. The problems in this final section tend to be accessible mainly through mental mathematics or an insight of some sort.

Problems have been placed in single classifications, though many could justifiably be placed in more than one section. In such instances, the classifications have been guided by the forms of the published solutions.

The fourteen sections of problems are followed by fourteen corresponding sections of solutions. Detailed solutions to all problems in the first thirteen sections are provided. Answers only are supplied for problems in the Quickies section, although a hint or insight is noted alongside the final answer for some of these problems.

At the back of the book, the sources of the problems are listed in the section entitled References. This section of the book has been divided into three parts to give proper credit to and facilitate readers' searches for sources. Contributors of the Problems offers an alphabetical list of all contributors' names. Each name is followed by a specification of the problem or problems with which the particular contributor is credited. Calendar Dates of the Problems offers an avenue for locating the original issue of the *Mathematics Teacher* in which a particular problem appeared. The calendar date corresponding to each problem is listed. The bibliography includes the references for any books that were identified by the contributors as sources of the problems.

The book includes three appendixes. Appendix A features solutions to the January 1995 calendar that is presented at the beginning of the book. Appendix B contains a list of all books that contributors noted as sources. In addition, some suggested readings for interested problem solvers are included in this appendix. Appendix C features the names of all individuals who served as calendar reviewers when the problems were originally being considered for inclusion in the *Mathematics Teacher*.

January 1995:
A Sample Calendar

The following two pages reproduce the January Calendar (*Mathematics Teacher* 88, pp. 40–41) from 1995—coincidentally, the year of NCTM's 75th anniversary. This calendar has been included here to introduce readers who are unfamiliar with the *Mathematics Teacher* to the format that is standard in the journal.

Designed to hang vertically like a conventional wall calendar, each calendar occupies the centerfold of the issue in which it appears. Classroom teachers can thus easily remove calendars from the issues in which they are stapled and can readily display them on their classrooms' bulletin boards.

All calendars present the days of the month in four columns, with one problem for each day. The sample shown here differs from the calendars in the journal in size and color but is faithful to the original format in all other respects. The journal's pages permit the calendars to be slightly larger than our sample can be in the pages of this book, and calendars in the journal are printed in a color in addition to black.

Solutions to each month's calendar problems are printed on the pages that wrap around the calendar. Consequently, when teachers remove a calendar from the center of an issue, most of the problems' solutions come with it, on the backs of the calendar pages. Since the solutions typically require more space than the statements of the problems, they usually run to a third page of the journal.

Instead of enfolding our sample with solutions in the customary way, on the page that precedes the calendar and the two pages that follow it, we have given the solutions to its problems in the back of the book, in Appendix A.

JANUARY

1

A science teacher decides to welcome the new year by recalibrating a thermometer as shown.
If normal body temperature is 37.0°C or 98.6°F, what is it on this new scale?

95° water boils

19° water freezes

2

Lagrange's four-square theorem states that every positive integer is the sum of the squares of four integers. For example, $19 = 3^2 + 3^2 + 1^2 + 0^2$ and $95 = 7^2 + 6^2 + 3^2 + 1^2$. Express 1995 as the sum of four squares. Is the answer unique?

3

Circle three numbers so that each column and each row contains exactly one of the numbers. What is the product of the three numbers? Why?

1	3	19
5	15	95
7	21	133

4

Find 19 consecutive integers whose sum is 95.

5

How many pairs of nonnegative integers x and y are solutions of
$$\frac{x}{19} + \frac{y}{95} = 1?$$

6

Find integers x, y, and z if—
1. the solid has exactly one symmetry plane.
2. no edge is longer than 19.
3. the volume is 1995.

7

If each letter stands for a different digit 1 through 9, find the two possible values for E.

$$\frac{N \cdot I \cdot N \cdot E \cdot T \cdot E \cdot E \cdot N}{N \cdot I \cdot N \cdot E \cdot T \cdot Y \cdot F \cdot I \cdot V \cdot E} = \frac{O \cdot N \cdot E}{F \cdot I \cdot V \cdot E}$$

8

If one side of a triangle is 19 and the perimeter is 95, what is the maximum possible area of the triangle?

9

A rectangle with area 19 is cut from the corner of a rectangle with area 95. Construct a line that separates the remaining region into two parts with equal areas.

10

A speed of 1995 cm per second is typical of a car being driven in which one of the following?

(a) In city traffic
(b) On a rural road
(c) On a superhighway
(d) In the Indianapolis 500 race

11

In what row and column is 1995?

1	3	6	10	15	...
2	5	9	14	20	...
4	8	13	19	26	...
7	12	18	25	33	...
11	17	24	32	41	...
16	23	31	40	50	...
...	...	...	...	...	

12

Find the ratio of the areas of the two equilateral triangles.

13

Today is Friday the 13th of January, 1995. In what year will 13 January next fall on a Friday?

14

Start at the point (19, 95), moving left, up, and diagonally at an angle as shown, repeating this movement until the y-axis is reached. What is the length of the path? (Give the answer in the form $a + b\sqrt{c}$.)

15

In a controversial school vote, only 19 percent of the teachers voted yes, whereas 95 percent of the students voted yes. If all teachers and students voted and 91 percent of the voters voted yes, what is the student-to-teacher ratio?

west and repeats this movement continuously. The first leg of its journey is 190 units, and each succeeding leg is half as long as the previous one. If the starting point is (−57, −57), what is the ultimate destination point?

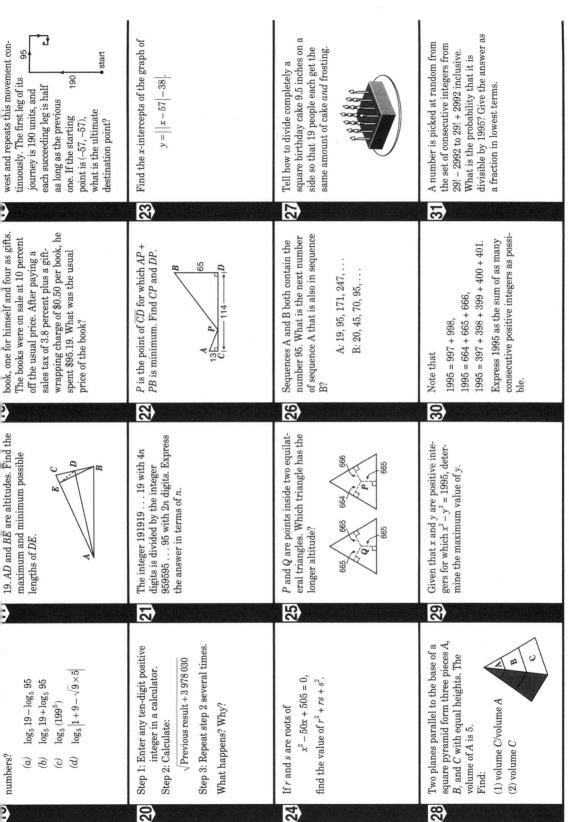

23

Find the x-intercepts of the graph of

$$y = \big|\,|x - 57| - 38\,\big|.$$

27

Tell how to divide completely a square birthday cake 9.5 inches on a side so that 19 people each get the same amount of cake *and* frosting.

31

A number is picked at random from the set of consecutive integers from $29! - 2992$ to $29! + 2992$ inclusive. What is the probability that it is divisible by 1995? Give the answer as a fraction in lowest terms.

---

book, one for himself and four as gifts. The books were on sale at 10 percent off the usual price. After paying a sales tax of 3.8 percent plus a gift-wrapping charge of $0.50 per book, he spent $95.19. What was the usual price of the book?

22

P is the point of CD for which $AP + PB$ is minimum. Find CP and DP.

26

Sequences A and B both contain the number 95. What is the next number of sequence A that is also in sequence B?

A: 19, 95, 171, 247, ...

B: 20, 45, 70, 95, ...

30

Note that

$1995 = 997 + 998$,

$1995 = 664 + 665 + 666$,

$1995 = 397 + 398 + 399 + 400 + 401$.

Express 1995 as the sum of as many consecutive positive integers as possible.

---

19. AD and BE are altitudes. Find the maximum and minimum possible lengths of DE.

21

Step 1: Enter any ten-digit positive integer in a calculator.

Step 2: Calculate:

$$\sqrt{\text{Previous result} + 3\,978\,030}$$

Step 3: Repeat step 2 several times.

What happens? Why?

25

P and Q are points inside two equilateral triangles. Which triangle has the longer altitude?

29

Given that x and y are positive integers for which $x^2 - y^2 = 1995$, determine the maximum value of y.

---

numbers?

(a) $\log_5 19 - \log_5 95$

(b) $\log_5 19 + \log_5 95$

(c) $\log_5 (199^5)$

(d) $\log_5 \big|\,1 + 9 - \sqrt{9 \times 5}\,\big|$

20

24

If r and s are roots of

$$x^2 - 50x + 505 = 0,$$

find the value of $r^2 + rs + s^2$.

28

Two planes parallel to the base of a square pyramid form three pieces A, B, and C with equal heights. The volume of A is 5. Find:

(1) volume C/volume A

(2) volume C

Readers' Favorites

The problems that appear in each calendar in the *Mathematics Teacher* feature solutions that are printed in the pages of the same issue. The publication of such solutions fortunately does not signify closure on the problems themselves. Rather, astute readers and solvers frequently offer alternate solutions, insights, or comments (sometimes corrections) on the problems and solutions. As the editor, I enjoyed reading the correspondence on these matters. Many of the letters appear in Reader Reflections, a regular feature of the *Mathematics Teacher*.

On reviewing hundreds of problems as an editor and later as the author of this book, I decided to draw attention here to some problems that took on enhanced significance as a result of the contributions offered by various problem solvers. Specifically, I will focus on five problems—namely, those for 5 December 1994, 29 November 1996, 12 April 1997, 31 May 1997, and 31 January 1998. Four of the five problems are from my tenure as editor. My selection of these problems is largely based on my own insight into them—insight that has indeed been enriched through correspondence and conversations! Brief considerations of these problems, including statements of the problems themselves, are offered here. I present the problems in chronological order and urge readers who find the discussions of any of them to be particularly interesting to refer to the original submissions published in the *Mathematics Teacher*.

5 December 1994

Construct a 75-degree angle with a compass and a straight edge.

—*Contributed by William Jamski*

This problem gave rise to a range of clever solution methods. The two published solutions (*Mathematics Teacher* 87, December 1994, p. 712) provided a basis on which other problem solvers built through contributions to Reader Reflections. Kenneth M. Zerone and Karil Johnson offered different approaches in the April 1995 issue of the journal (p. 358). Zerone's solution required only eight arcs to be constructed, and Johnson's employed constructions of equilateral triangles and multiple bisections of angles. Several months later, my personal preference among the methods appeared, in the October 1995 issue of the *Mathematics Teacher* (p. 622). This solution, proposed by Daniel B. Hirschhorn, was based on the construction of circles (see fig. 1). The beauty of the technique is enhanced by the fact that the compass width remains fixed throughout the entire process. Moreover, the resulting diagram is extremely appealing visually, since the circles give a sense of completion, in a picture that is unmarred by arc marks and segments that might seem extraneous to the untrained eye.

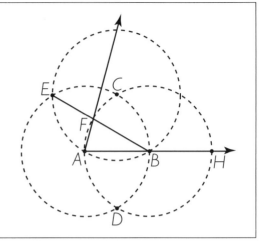

Suppose $\overrightarrow{AB}$ is given and that the goal is to construct an angle with vertex at A whose measure is 75 degrees. Set the compass length at radius AB. Draw circles centered at A and B. These circles intersect at C. Draw a circle centered at C with radius AB. Circle C and circle A intersect at E. $\overline{BE}$ and circle B intersect at F. $\angle FAB$ is the desired angle.

Proof. $\triangle ABC$ is equilateral, so $m\angle ABC = 60°$. $\overline{BE}$ bisects this angle, so $m\overarc{AF} = 30°$. This result makes $m\overarc{FH} = 150°$. Thus $m\angle FAB = 1/2 \cdot m\overarc{FH} = 75°$.

Fig. 1. Daniel B. Hirschhorn's solution to problem 5, December 1994

29 November 1996

A narrow slice cut from a pie with radius ten inches is to be divided in half by cutting it as shown. How far from the top should the cut be made so that the two pieces will be of approximately the same area?

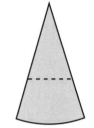

(*a*) About 6 inches (*b*) About 7 inches

(*c*) About 8 inches (*d*) About 9 inches.

—*Contributed by Gene Zirkel*

In a letter that appeared in Reader Reflections in May 1997 (p. 415), James E. Kessler offered a simpler approximation than that in the published calendar solution (*Mathematics Teacher* 89, November 1996, p. 658). Instead of approximating the slice with a triangle, he approached it as a sector of a circle. Kessler's letter in turn sparked the interest of John Goehl and J. G. Roof (*Mathematics Teacher* 91, January 1998, pp. 73–74), both of whom prepared detailed discussions that added considerably to the value of the problem as a teaching example. Goehl's letter was based on the premise that exactness would not be much harder to obtain (see fig. 2). Meanwhile, Roof chose "to do a rigorous treatment on more generous-sized slices" (p. 73) rather than an approximation using narrow slices. Both solutions depended heavily on trigonometry.

12 April 1997

Arrange the integers from 1 through 20 in a line so that the sum of each adjacent pair is a prime number. For example, a prime line of length 4 is 1 2 3 4. Note that $1 + 2 = 3$, $2 + 3 = 5$, and $3 + 4 = 7$.

—*Contributed by Margaret J. Kenney and Stanley J. Bezuszka*

This problem generated more correspondence than can be presented in a reasonable space. Some of the correspondence took place "off the pages." Doug Rogers directed numerous e-mails my way concerning the problem. His examination of it has become quite extensive, in fact. A snapshot of his work is presented in the second letter that I discuss here. However, it was James W. Roche who deserves much of the credit for engaging

The original solution approximates the sector with a triangle. Kessler approximates the triangle with a sector. The exact solution to the problem requires that

$$\frac{1}{2}bh = \frac{1}{2}\left(\frac{1}{2}\theta r^2\right).$$

In this equation, b and h are the base and height of the triangle, and θ and r are the angle (in radians) and radius of the sector.

Let x be the radius of Kessler's sector. Then

$$x = \frac{h}{\cos\left(\frac{1}{2}\theta\right)}.$$

Since

$$b = 2x \sin\left(\frac{1}{2}\theta\right),$$

$$bh = 2x^2 \sin\left(\frac{1}{2}\theta\right)\cos\left(\frac{1}{2}\theta\right)$$

$$= x^2 \sin\theta = \frac{1}{2}\theta r^2.$$

Hence, the exact solution is

$$x = \left(\frac{\frac{1}{2}\theta}{\sin\theta}\right)^{\frac{1}{2}} r.$$

As $\theta \to 0$,

$$x \to \frac{r}{\sqrt{2}} \approx 0.7071 lr.$$

Thus, Kessler's result is correct in this limit. Values of x for some other angles are shown in table 1.

Notice the interesting situation for large angles. As θ increases, x eventually becomes greater than the radius of the sector. The angle at which this result occurs is found by setting $x = r$, which implies that $(1/2)\theta = \sin\theta$. The value of θ can be found numerically using Newton's method. Take $f(\theta) = \sin\theta - (1/2)\theta$. Then

$$\theta_{n+1} = \theta_n - \frac{f(\theta_n)}{f'(\theta_n)}$$

$$= \theta_n - \frac{\sin\theta_n - \frac{1}{2}\theta_n}{\cos\theta_n - \frac{1}{2}}.$$

Converting the result from radians to degrees yields $\theta \approx 108.6°$.

The figure shows the situation for θ greater than $108.6°$.

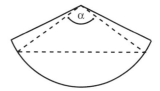

Since the segment has half the area of the sector,

$$\frac{1}{2}r^2(\alpha - \sin\alpha) = \frac{1}{2}\left(\frac{1}{2}\theta r^2\right)$$

or

$$\alpha - \sin\alpha = \frac{1}{2}\theta,$$

where α is the angle of the segment. The distance from the tip is found from $r\cos(1/2 \cdot \alpha)$. Returning to the case of $\theta = 120°$, this distance is found from the zeros of the function

$$f(\alpha) = \alpha - \sin\alpha - 60\pi/180.$$

Again applying Newton's method gives

$$\alpha_{n+1} = \alpha_n - \frac{f(\alpha_n)}{f'(\alpha_n)}$$

$$= \alpha_n - \frac{\alpha_n - \sin\alpha_n - \frac{\pi}{3}}{1 - \cos\alpha_n}.$$

Converting the result from radians to degrees yields $\alpha \approx 112.8°$ and $r\cos(1/2 \cdot \alpha) \approx 0.553r$ as the distance from the tip.

Table 1

Values of x for Various Angles

θ (degrees)	θ (Radians)	x/r
30	$\frac{\pi}{6}$	$\left(\frac{\pi}{6}\right)^{\frac{1}{2}} \approx 0.7236$
60	$\frac{\pi}{3}$	$\left(\frac{\pi\sqrt{3}}{9}\right)^{\frac{1}{2}} \approx 0.7776$
90	$\frac{\pi}{2}$	$\left(\frac{\pi}{4}\right)^{\frac{1}{2}} \approx 0.8862$
120	$\frac{2\pi}{3}$	$\left(\frac{2\pi\sqrt{3}}{9}\right)^{\frac{1}{2}} \approx 1.0996$

Fig. 2. John Goehl's solution to problem 29, November 1996

readers in thinking about this problem. My own interest stems from his thoughtful letter, which appeared in Reader Reflections in February 1998 (p. 155):

> I was especially intrigued with "Calendar" problem 12 of April 1997.... I kept getting hung up at the end until I discovered that if I start with the largest number and then put next to it the largest allowable number and continue with this method, the technique seems to work.
>
> For $n = 20$:   20, 17, 14, 15, 16, 13, 18, 19, 12, 11, 8, 9, 10, 7, 6, 5, 2, 3, 4, 1.
>
> For $n = 30$:   30, 29, 24, 23, 20, 27, 26, 21, 22, 25, 28, 19, 18, 13, 16, 15, 14, 17, 12, 11, 8, 9, 10, 7, 6, 5, 2, 3, 4, 1.
>
> I pose the question: does this method work for any n? I have not found a counterexample, but I have not been able to prove my method in general. Perhaps some readers will be able to prove or disprove my "method."
>   This problem prompted excellent discussion in my classes.

I enjoyed playing with the conjecture posed by Roche, and my own response to Roche's letter was published simultaneously (*Mathematics Teacher* 91, February 1998, pp. 155, 183:

> The method of starting with the largest number worked for all the examples that I tried. The question of whether the method always works remains open. I would like to offer a variation on Roche's idea. I shall assume that the method does work for all values of n greater than 1. (A counterexample is welcomed!) I wish to strive for an efficient way of showing that all values of n from 2 up to some arbitrary value can be produced. We do not need to consider each value of n separately; we can avoid unnecessary work. Let me explain by way of Roche's own example for $n = 20$.
>   Observe that the string for $n = 20$ begins 20, 17, 14, 15, 16, 13, 18, 19, ..., thus leaving us with the integers from 1 to 12 as the final twelve numbers. The fact that 12 appears next assures us that an arrangement for $n = 12$ exists. Similarly, a solution for $n = 11$ is nested in the arrangement for $n = 20$. In fact, Roche's arrangement proves that arrangements exist for each of $n = 20$, 12, 11, 7, 6, and 5. Suppose that we wanted to show that all values from 2 to 20 are possible to obtain. The largest-number technique could be modified to consider $n = 19$. We would get the following arrangement:
>
> $$19, 18, 13, 16, 15, 14, 17, 12 \ldots.$$
>
> We may stop upon reaching 12, because we already know what the final sequence of digits will look like. Note that we progressed toward our objective by showing that both $n = 19$ and $n = 18$ produce satisfactory arrangements. For $n = 17$, we get
>
> $$17, 14, 15, 16, 13, 10, 9, 8, 11, 12, 7 \ldots.$$
>
> This example added $n = 17$ and $n = 13$ to our list of verified values. Note that $n = 7$ has already been verified, thus making it unnecessary to proceed further with this string. Starting with $n = 16$, we get an arrangement that verifies the method for $n = 16$, 15, 14, and 2.
>
> $$16 \ 15 \ 14 \ 9 \ 10 \ 13 \ 6 \ 11 \ 12 \ 7 \ 4 \ 3 \ 8 \ 5 \ 2 \ 1$$
>
> Only five values of n remain to be verified—namely, 10, 9, 8, 4 and 3. Starting with $n = 10$, we get
>
> $$10 \ 9 \ 8 \ 5 \ 6 \ 7 \ 4 \ 3 \ 2 \ 1.$$
>
> Aha! This sequence verifies the five remaining values.
>   In summary, five trials were required, beginning with 20, 19, 17, 16, and 10, to verify that all values from 2 to 20 could be obtained. Suppose that we select $n = 100$ or 47 or 251 or ... [any other value]. How many trials would be required? I expect that more than five trials

would be required. A comparison of raw numbers does not interest me here. I would like to compare the relative efficiencies associated with the process and different values of n.

Suppose that we define an efficiency index as follows:

efficiency index = number of trials/ (largest) value of n

For the example discussed, the efficiency index would equal 5/20, or 0.25. Ironically, a lower efficiency index indicates a more efficient process. I challenge readers to determine which numbers produce the lowest efficiency indexes. For practical purposes, the challenge becomes more interesting if we consider only values of n greater than 15. The use of computers would make it practical to consider very large values of n; however, the challenge to find the most efficient value of n between 15 and 100 would be reasonable without a computer. Remember to keep your worked examples handy. You can then identify strings of digits that have been used in previous trials, thus reducing the need for unnecessary work. An inspection of the examples may suggest starting points for further investigation.

As mentioned previously, Doug Rogers provided me with considerable feedback on this problem, including concerns about the efficiency index that I proposed in my response to Roche. Rogers provided some relevant historical information in the following personal correspondence with me:

One of your proposers of this problem, M. J. Kenney, also offered a version of this problem in the *Student Math Notes* accompanying the November 1986 issue of the *NCTM News Bulletin*. Richard Guy has written about it in *Crux Mathematicorum* 19 (1993), 97–99, raising rather more questions than are answered.

The algorithm proposed more recently in the correspondence section of your February 1998 issue by J. W. Roche is what might be termed the "greedy" algorithm, being what a computer program might try, before resorting back to any backtracks. It does seem to hold up for quite large starting values (for example, all those between 600 and 630), but there is a suspicion that it should not, because the constraints towards the end might seem too exacting to satisfy.

My belief is that this problem, in terms of its breadth and potential, proved to be the richest one that I handled during my editorial tenure. The problem is extremely accessible. Clearly, it offers a very fertile field for exploration and investigation of algorithms, conjectures, and proofs. The references provided by Rogers confirm that this is indeed a rich problem. Before proceeding to read Guy's earlier article, I expect to play a little more with some of the underlying ideas expressed through Reader Reflections and the correspondence from Doug Rogers and, subsequently, Monte Zerger. I am indebted to James W. Roche for initiating the discovery of the wonders associated with a seemingly elementary challenge.

31 May 1997

Find the numbers A, B, C, D, E, and F for the six sectors so that the numbers in the sectors, together with totals from sets of adjacent sectors, give all the integers from 1 to 27, inclusive.

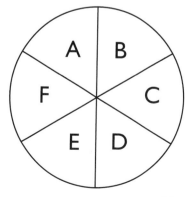

Shortly after the publication of this problem and the accompanying solution (1, 2, 5, 9, 6, 4), Mike Diehl, a high school student, offered a new solution that appeared in Reader Reflections (*Mathematics Teacher* 90, November 1997, p. 683). His sequence, 1, 6, 2, 5, 11, 3, was different from that in the published calendar solution (*Mathematics*

Teacher 90, May 1997, p. 379). I noticed that the sum of his numbers was 28, a fact that extended the value of his solution beyond that of an alternate set of numbers and made it worthy of publication. A subsequent letter from Vikash Gilja and Julie Hong of Manhasset (N.Y.) High School also earned a place in Reader Reflections (*Mathematics Teacher* 91, May 1998, pp. 396–97). Gilja and Hong presented thirty-nine additional solutions, five of which actually provide all the integers from 1 to 31, inclusive. For example, 1, 2, 5, 4, 6, 13 is one such arrangement. The others appear in the extensive list furnished in the letter from Gilja and Hong.

31 January 1998

Divide this checkerboard along the edges of the squares to get four congruent pieces, each containing one of the circles.

—*Contributed by Victor G. Feser*
(*adapted from* Mathematical Puzzles of Sam Loyd *[1959]*)

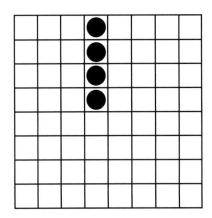

Various configurations, in addition to the published solution (*Mathematics Teacher* 91, January 1998, p. 60; shown here as fig. 3a), appeared in Reader Reflections in the October 1998 issue of the *Mathematics Teacher* (p. 626). Alternate solutions provided by Ben Nahir and Eli Bogart (see fig. 3b) and Helena Rocha sparked responses from the problem's contributor, Victor Feser. His replies were published simultaneously with the new solutions in Reader Reflections. Feser congratulated Nahir and Bogart, who were then high school students, on a solution that suggested a pattern that in turn led to two additional solutions (shown in fig 3c). He also commended Helena Rocha, noting that her solution (fig. 3d), together with the others shown in figure 3, now gave five solutions to the original problem. Feser invited readers to think further about the problem:

Do still more solutions exist? One possibility to explore: In each of the solutions, the four congruent pieces can be made to coincide by a simple *rotation*—they form a sort of pinwheel. (*Mathematics Teacher* 91, October 1998, p. 626)

In the lively spirit of problem solving, Feser challenged readers to determine whether a solution requiring a *reflection* exists.

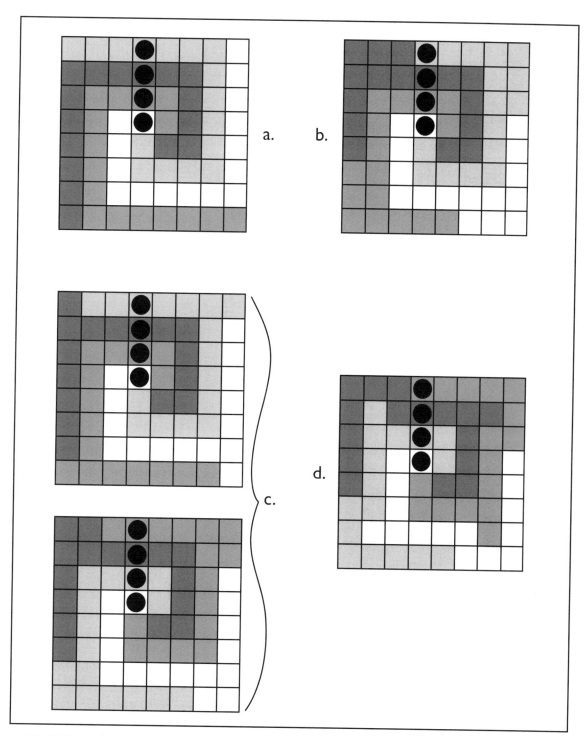

Fig. 3. Five solutions to problem 31, January 1998: *a*, published solution; *b*, alternate proposed by Ben Nahir and Eli Bogart; *c*, two more solutions from problem contributor Victor Feser; *d*, a fifth solution offered by Helena Rocha

Problems

Number Theory

$5^2 - 1 = 24$ 8

1. Select any prime number greater than 3. Square it and subtract 1. What is the largest number that must be a divisor of the result? $7^2 - 1 = 48$

2. The number 75 may be written as the sum of two consecutive whole numbers, $37 + 38$, and as the sum of three consecutive whole numbers, $24 + 25 + 26$. Write 75 as the sum of other consecutive whole numbers and as the sum of consecutive integers.

3. A number is *deficient* if it is greater than the sum of its proper factors. Explain why a prime number must be a deficient number.

4. A number is called *cute* if it has exactly four positive integer divisors. What percent of the first twenty-five positive integers are cute?

5. Determine the largest prime divisor of $87! + 88!$.

6. A peddler is taking eggs to the market to sell. The eggs are in a cart that holds up to 500 eggs. If the eggs are removed from the cart 2, 3, 4, 5, or 6 at a time, one egg is always left over. If the eggs are removed 7 at a time, no eggs are left over. How many eggs are in the cart?

7. Given the following clues, determine my telephone number: The first three digits appear in descending order with a sum of 18 and a product of 210. The final four digits consist of three primes in ascending order surrounding a composite number in the second-to-last position. These four digits have a product of 336.

8. Some of the factors of a locker number are 2, 5, and 9. If it has exactly nine additional factors, what is the locker number?

9. If a 75-degree sector rotates around the center of a circle in 75-degree movements, how many sector movements will be needed for the sector to land directly in its original position?

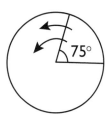

10. What is the remainder when 2^{151} is divided by 7?

11. Find the units digit of the decimal numeral representing the number $11^{11} + 14^{14} + 16^{16}$.

12. How many four-digit numbers that end in 75 are divisible by 75?

13. Which pairs of natural numbers have squares that differ by 75?

14. Does 124_{five} represent an odd number? How can you determine whether a number is odd by looking at its base-five representation?

15. A four-digit palindrome can be represented by $abba$. Show that this expression is always divisible by 11.

16. A certain three-digit number in base ten with no repeated digits can be expressed in base R by reversing the digits. Find the smallest value of R.

17. Adjoin to the digits 739 three more digits so that the resulting number 739… is divisible by 6, 7, 8, and 9.

18. The decimal numeral $1,287,xy6$, where x and y stand for decimal digits, represents a multiple of 72. Find all the ordered pairs (x, y).

19. What are the final two digits of 7^{1997}?

20. Find the prime factors of 1,000,001.

21. Prove that the sum of three consecutive integers is divisible by 3, that the sum of five such integers is divisible by 5, but that the sum of four consecutive integers is not divisible by 4.

22. For what base value B will 1996_B equal $32,225_{\text{ten}}$?

23. In the set $X = \{1, 2, 3, …, 600\}$, the elements that are multiples of 3 or 4, or both, are assigned to a subset Y. Find the sum of all the elements of Y.

24. Find the largest value of n for which 2^n will be a factor of 20!.

25. Arrange the digits 0 through 9 so that the first digit is divisible by 1, the first two digits are divisible by 2, and the first three digits are divisible by 3. Continue until the first nine digits are divisible by 9 and the ten-digit number is divisible by 10.

26. Find five positive integers such that the greatest common divisor of any two is equal to their difference.

27. Find all integers with an initial digit 6 having the property that if the initial digit is deleted, the resulting number is exactly 1/25 of its original value.

28. When an integer n is divided by 1995, the remainder is 75. What is the remainder when n is divided by 57?

29. Find the values of A, N, and E given that $ANNE_{\text{eight}} - ANNE_{\text{five}} = ANNE_{\text{seven}}$.

30. We can express 7/8 as a decimal, namely, .875. When written in the form

$$\frac{7}{8} = \frac{1}{2} + \frac{1}{4} + \frac{1}{8},$$

we can express 7/8 as a bicimal (base two) in the form

$$\frac{7}{8} = .111_{(\text{two})}$$

Express 2/3 as a bicimal.

31. For which positive integers n is it true that $3n - 7$ divides $5n + 3$?

Coordinate Geometry

1. A point (x, y) with integral coordinates is called a lattice point. For example, $(-1, 0)$ is a lattice point, whereas $(3, 1/2)$ is not. How many lattice points lie on the graph of $x^2 + y^2 = 25$?

2. The areas of these two adjacent squares are 4 cm² and 196 cm². Find the length of the segment joining the centers of their inscribed circles.

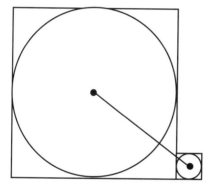

3. The two squares have dimensions as indicated. What is the area of the shaded triangle?

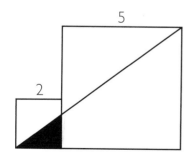

4. Three vertices of a parallelogram are $(1, 1)$, $(3, 5)$, and $(-1, 4)$. Find all possible ordered pairs that could be the coordinates of the fourth vertex.

5. Find the area of the region enclosed by the graph of $|y| + |2x| = 6$.

6. A, B, and C are the points $(1, 2)$, $(0, 0)$, and $(-1, 3)$, respectively. What is the slope of the bisector of $\angle ABC$?

7. $ABCD$ is a unit square with D at $(1, 0)$ and C at $(2, 0)$. Find the equation of the line through the origin that bisects the area of this square.

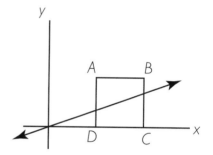

8. For the tower of nine squares shown, determine a line passing through point P that will split the area of the nine squares into two equal parts.

9. Find all points (x, y) that have an x-coordinate twice the y-coordinate and that lie on a circle of radius 5 with center at $(2, 6)$.

10. What is the area of the set of points (x, y) such that $|x| - |y| \leq 1$ and $|y| \leq 1$?

11. In the region $-1 < x < 1/2$, the graph of the function $y = |x + 1| + |2x - 1|$ coincides with the graph of the function $y = ax + b$. Find a and b.

12. A circle of radius 4 is centered at the origin; every second, its radius increases by 3 units. A second circle, of radius 12, is centered at $(30, 0)$; every second, its radius decreases by 1 unit. This process continues until the circles meet. At that time, the point $(27, 4)$ lies in which location?

   (a) inside the first circle, (b) on the first circle, (c) inside the second circle, (d) on the second circle, or (e) between the circles.

13. Regular octagon $ABCDEFGH$, with its vertices labeled in clockwise order, is drawn on a rectangular coordinate plane. The coordinates of A are $(4, 0)$ and the coordinates of B are $(0, 4)$. If the coordinates of vertex E are (p, q), compute $p - q$.

14. If a is randomly chosen from $M = \{-3, -2, -1, 0, 1, 2, 3\}$, b is randomly chosen from $N = \{-3, -2, -1, 0, 1, 2, 3\}$, and c is randomly chosen from $P = \{-6, -4, -2, 0, 2, 4, 6\}$, determine the probability that $(x, y) = (-1, 2)$ is a solution to $ax + by = c$.

Spatial Sense

_#1, 3, 4, 5, 6 7_

1. Three views of the same block are shown below. What letter is on the side parallel to the side with the letter A?

2. Add some squares to this figure so that it has exactly (a) one line of symmetry, (b) two lines of symmetry, and (c) no lines of symmetry.

3. If town A and town B are eight miles apart and town C is ten miles from town B, what are the possible distances from town A to town C?

4. Match the pattern with its folded figure.

5. Of the two figures, one can be drawn without lifting a pencil or retracing a line segment, but the other cannot. Which can be drawn in this manner?

 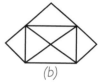

6. The figure shown consists of twenty unit squares. Divide this figure into five pieces, each consisting of four unit squares, so that no two pieces have the same shape.

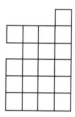

7. I have folded a piece of paper in half and then in half again. Next I cut a shape from the folded paper as shown.

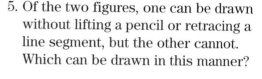

When I unfold the paper, it looks like which of the following?

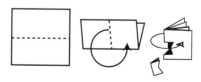

8. The following equilateral triangle has been dissected into four congruent equilateral triangles.

Using isometric dot paper, dissect an equilateral triangle into six congruent triangles.

9. Divide a rectangle into four similar triangular regions.

10. Eight squares of equal size are placed on a table. Square A is seen completely, but the other seven are overlapped and can be seen only partially. Find the order in which the squares are placed from top to bottom (i.e., placed last to first).

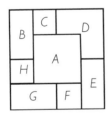

11. Arrange six points in a plane so that any three of the points chosen will form the vertices of an isosceles triangle.

12. Given the following figure with each segment of length 1, use exactly two cuts to divide it so that it can be arranged into a rectangle with dimensions in a ratio of 2:1.

13. All corners were sliced off a cube whose volume was 2000 cubic centimeters. Triangular faces are located where the cube's vertices were. Each triangle is equilateral with 5-centimeter sides. Determine the total number of edges on the new solid.

14. These two rectangular grids can be separated into congruent "stairsteps" by cutting steps along grid lines. What other rectangular grids can be cut into congruent stairsteps?

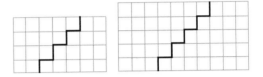

15. Draw a rhombus that has a pair of 60-degree angles. Decide where to place a mirror upright on the rhombus to produce an image of a hexagon.

16. Decide where to place a mirror upright on a square to produce an octagonal image.

17. A regular hexagon has been dissected into two congruent regions in two ways, as shown.

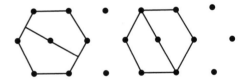

Using isometric dot paper, dissect a regular hexagon into twelve congruent regions in two ways.

18. Find all points in a plane that are 1 cm from a given segment 4 cm long and are also 2 cm from the midpoint of the segment.

19. Without retracing any part of your path, draw five connected line segments that pass through all twelve points. The line segments must form a closed path—that is, the last segment must end where the first segment began.

```
•   •   •   •

•   •   •   •

•   •   •   •
```

20. Without retracing any part of your path, draw six connected line segments that pass through all sixteen points. The line segments must form a closed path—that is, the last segment must end where the first segment began.

```
•   •   •   •

•   •   •   •

•   •   •   •

•   •   •   •
```

21. Consider a 12×12 chessboard consisting of 144 1×1 squares. If I can remove 3 corner squares, can I cover the remaining 141 squares with 47 1×3 tiles?

Logic

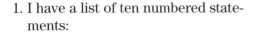

1. I have a list of ten numbered statements:

   1. Statement 2 is false.
   2. Statement 3 is false.

      .
      .
      .

   9. Statement 10 is false.
   10. Statement 1 is false.

   How many statements are true? How many are false?

2. You have three boxes: a red box, a blue box, and a green box. Inside each box is a single ball of the same color as the box. Someone takes each ball out and puts it in a different box. You open the red box and discover a green ball. What color ball is in the blue box?

3. Read the following three statements and determine which are true and which are false:

   1. There are three numbered statements.
   2. Two of the numbered statements are not true.
   3. This statement is true.

4. You are shown the following four cards. Each card has a single positive integer 1, 2, 3, or 4 on each side, but you can see only one side of each card. Some numbers may appear more than once. How many cards must you turn

over to verify that any card that has a 2 on one side also has a 4 on the opposite side?

5. During the Depression, two women each had 30 apples to sell. One sold hers at 2 for a nickel; the other, at 3 for a nickel. They ended the day with 75 cents and 50 cents, respectively, or $1.25 in all. The next day, they decided to pool their apples and sell them at 5 for a dime (2 for a nickel and 3 for a nickel). At day's end they had only $1.20. A dispute arose over the missing nickel. What happened to it?

6. A man and a woman were the only two people walking along a beach. "I'm a man," said the one with black hair. "I'm a woman," said the one with blonde hair. At least one of them was lying. What was the color of the woman's hair?

7. Two clubs have a Halloween initiation. Those joining the Spooks tell the truth. Those joining the Goblins always lie. Three initiates approached a door to trick or treat. A woman said, "What club are you joining?" The first boy's response was unclear, but the second said, "He said he was joining the Spooks." The third said, "You're lying!" What club was the third initiate joining?

24

8. John does not know the pairings of the final eight teams remaining in the NCAA tournament. Only four will advance to the finals, and he hopes that his WKU team is one of them. He finds that the *Nashville Banner* has picked UCLA, UT, OU, and UK to advance; the *Courier-Journal* has picked UK, OU, MSU, and UNCC; and the *Daily News* has picked UK, UT, MSU, and OSU. Determine the pairings of the final eight.

9. Ten teams play in a tournament. Each game involves two teams. If the tournament consists of twenty-three games, explain why some team must play five or more games.

10. Thirty-five students are seated in five rows and seven columns. Is it possible for the students to change seats if every student must move exactly one seat to the left, right, front, or back?

11. If the ten digits {0, 1, 2, …, 9} are arranged in any order on a circle, show that three consecutive digits always appear whose sum is at least 15.

12. "Mr. Steuben has a hundred dollars or more in his wallet," said Paula. "Mr. Steuben has less than a hundred dollars in his wallet," said Pauleen. "Well, I know he has some money in his wallet," said Paulette. If just one of these statements is true, then how much money does Mr. Steuben have in his wallet?

13. Myrtle has two white balls, two black balls, and two boxes. She may place the balls in the boxes in any way that she pleases. Her husband, Max, will then pick a box without looking into it, and with his eyes closed, pick out a ball. If he draws a white ball, the couple wins $500. How should Myrtle arrange the balls to maximize the probability of winning?

14. Those feline mathematician-explorers Zoe and Reginald are captured by a fearsome Sphinx who holds two checkered squares: one is 8×8, and the other is 6×6. She says, "Cut them both into two pieces and out of the four resulting sections form a 10×10 square. If you can, I will release you." How can they escape from this deadly trap?

Algebra

3, 4, 5

1. Find the sum of the reciprocals of two numbers, given that these numbers have a sum of 50 and a product of 25.

2. Solve for N:

$$6! \ 7! = N!$$

3. Rather than use the cumbersome $F = 9/5 \ C + 32$ for converting from Celsius to Fahrenheit, could you instead mentally double C and subtract 10 percent of the result before adding 32? Why or why not?

4. Label the figure to illustrate the identity

$$4ab + (a - b)^2 = (a + b)^2.$$

5. The average (arithmetic mean) of a and b is 10, whereas the average of b and 10 is $c/2$. What is the average of a and c?

6. Define the operation ♦ as follows:

$$a \diamond b = \text{GCD} \ (a, b) + \text{LCM} \ (a, b)$$

For example, $6 \diamond 8 = 2 + 24 = 26$. Show that the operation ♦ is not associative.

7. The sum of two numbers is 45. The sum of their quotient and its reciprocal is 2.05. Find the product of these two numbers.

8. $3x + 7y = 188$. What integer solution (x, y) yields the smallest positive difference $(y - x)$?

9. For 1975 through 1990, the number of master's degrees, M, awarded in library and archival science in the United States can be modeled by

$$M = 670\sqrt{280 - 31.6x + x^2},$$

where $x = 5$ represents 1975. In which year prior to 1990 were approximately 5360 masters degrees earned?

10. Given that $K^2 - 3K + 5 = 0$, determine the value of $K^4 - 6K^3 + 9K^2 - 7$.

11. What does

$$\sqrt{6 + \sqrt{6 + \sqrt{6 + \sqrt{6 + \ldots}}}}$$

equal?

12. The planes defined by the equations $6x + y + 2z = -1$ and $x + 2y = 0$ intersect at the point $(k, k^2, k^3 + 1)$. Find k.

13. Find the values of x and y for which $(x + yi) = (1 + i)^{12}$ where $i^2 = -1$.

14. A three-digit number grows by 9 if we exchange the second and third digits and grows by 90 if we exchange the first and second digits. By how much will it grow if we exchange the first and third digits?

15. Using the balance-scale pictures shown, tell how many ◆s are necessary to replace the ? in the last picture.

16. If $(mx + 7)(5x + n) = px^2 + 15x + 14$, find $m(n + p)$.

17. Find the product of all real values of p that satisfy the equation

$$6|p - 6| = |p + 6|.$$

18. If $m + n = 3$ and $m^2 + n^2 = 6$, find the numerical value for $m^3 + n^3$.

19. The sum of two of the three roots of $x^3 + ax^2 + bx + c = 0$ is 0. What equation expresses c explicitly in terms of a and b?

20. Show that the quartic equation

$$(\ )x^4 + (\ )x^3 + (\ )x^2 + (\ )x + (\ ) = 0$$

always has a rational root when the numbers 1, –2, 3, 4, and –6 are randomly assigned to fill the five parentheses.

21. For what value(s) of the coefficient a do the equations

$$x^2 - ax + 1 = 0$$

and

$$x^2 - x + a = 0$$

have a common real solution?

22. The *adderage* of two fractions is defined as the sum of their numerators over the sum of their denominators. When, if at all, will the adderage of two fractions equal the average of the same two fractions?

23. The numbers a, b, c, and d are such that the sum of any one of them and the product of the other three is equal to 2. If $a > b$, $a > c$, and $a > d$, find a.

24. Heron's formula states that the area of a triangle with sides of lengths a, b, and c equals

$$\sqrt{s(s - a)(s - b)(s - c)}$$

where $s = (1/2)(a + b + c)$. Verify Heron's formula for a right triangle with sides of lengths a, b, and c such that $a^2 + b^2 = c^2$.

25. Consider the following magic square:

A	B	C
D	E	F
G	H	I

Show that

$$E = \frac{1}{9}(A + B + C + \ldots + I).$$

26. The incorrect cancellation of the sixes in the following leads to a correct answer:

$$\frac{16}{64} = \frac{1}{4}.$$

For how many other values of digits A, B, and C is it true that

$$\frac{AB}{BC} = \frac{A}{C},$$

where $AB = 10A + B$ and $BC = 10B + C$?

27. Find (a, b, c, d) if $[a, b, c, d] = 5/13$ and $a, b, c,$ and d are positive integers such that

$$[a,b,c,d] = \cfrac{1}{a + \cfrac{1}{b + \cfrac{1}{c + \cfrac{1}{d}}}}.$$

28. Herr Vitami buys three kinds of fruit for 40 schillings, 10 schillings, and 1 schilling each. He pays 259 schillings for 100 pieces of fruit. How many pieces of the cheapest kind of fruit did he buy?

29. The roots of $x^2 + bx + c = 0$ are r_1 and r_2 where $|r_1 - r_2| = 1$. Express c in terms of b.

30. A right circular cylinder has been sliced into pieces perpendicular to the axis. The first piece is 1/2 the original cylinder. The second piece is 1/3 of the remaining half, the third is 1/4 of the remainder, and continuing on as long as possible. (*a*) Describe algebraically the *n*th term of the sequence. (*b*) Describe algebraically the partial sum of *n* terms.

31. For all real x, $f(x) = x + 3$ and $g(x)$ is a polynomial of degree 2 such that $g(f(x)) = x^2 + 2$. Find $g(x)$.

32. If $f(x)$, defined for all real x, satisfies $f(1 - x) + 2f(x) = x$, express $f(x)$ as a polynomial in x.

33. Given that m and n are integers such that $1/3 < m/n < 1$, find all values of m/n that do not change when a particular integer is added to the numerator and simultaneously multiplied by the denominator.

34. Find the smallest positive number that leaves remainders of 3, 4, 5, and 6 when divided by 8, 9, 11, and 13, respectively.

35. Find the maximum value of x^2/y^2 given that $5x^2 - 12xy - 18y^2 = 0$, where x and y are nonzero real numbers. Write the value in the form

$$\frac{a + b\sqrt{c}}{d},$$

where a, b, c, and d are integers.

Probability

1. A golf ball falls randomly onto a circular green 10 meters in radius, with the cup at the center. What is the probability that the ball falls within 1 meter of the cup?

2. A pair of six-sided dice is tossed. What is the probability that the pip total is a prime number?

3. Three vertices of a regular hexagon are randomly selected. What is the probability that an isosceles triangle would be formed by connecting these three vertices?

4. What is the probability that a number chosen at random from the range 1,000,000,000 to 9,999,999,999 inclusive will contain ten different digits? (Round your answer to the nearest millionth.)

5. Three coins are tossed. Given that at least one coin lands heads up, determine the probability that all three coins land heads up.

6. A box contains four digits: one 1, two 9s, and one 8. The digits are written on separate pieces of paper. The digits are drawn without replacement and placed in the order they are drawn to create a four-digit number. What is the probability that the four-digit number is prime?

7. If two marbles are removed at random from a bag containing black and white marbles, the chance that they are both white is 1/3. If three are removed at random, the chance that they are all white is 1/6. How many marbles of each color are in the bag?

8. A positive integer, n, is picked at random. What is the probability that the greatest common factor of n and 30 is greater than 1? Express your answer as a fraction in lowest terms.

9. A number is randomly selected from 1 through 100. Given that the number selected is prime, what is the probability that it contains the digit 9?

10. Two players flip three fair coins each. What is the probability that they get the same number of heads?

11. Three dice are thrown. What is the probability that the three dice will not all show the same number?

12. A die is to be rolled six times. How many times as likely is the probability that each face will come up exactly once as is the probability that a 6 will come up on each throw?

13. Anna says that a 55 percent chance exists that she will go to a movie tomorrow if it is raining at noon and a 30 percent chance if it is not raining at noon. Willard forecasts a 40 percent chance of rain at noon. On the basis of these numbers, what is the probability that Anna will go to a movie?

14. If a six–sided die numbered 0 through 5 is rolled until the sum of the numbers rolled is greater than 12, what sum is most likely?

15. Three dart players threw simultaneously at a ticktacktoe board, each hitting a different square. What is the probability that the three hits constituted a win at ticktacktoe?

16. Given that a number is a three-digit palindrome, what is the probability that it is divisible by 11?

17. Three players take turns—player A first, then B, then C, then A, and so forth—rolling a die. The first to roll a 6 is the winner. What is each player's probability of winning?

18. A basketball player attempts a free throw. If she is successful, she gets to attempt a second free throw. If p is her probability of a successful free throw, and if her probability of making zero points is equal to that of making two points, what is p? (Round the answer to the nearest thousandth.)

19. A box contains five distinct rods measuring 15, 30, 40, 60, and 90 centimeters in length. Three of these rods are randomly chosen. What is the probability that they can be arranged to form a triangle?

20. In Indianapolis, the distance between the conference center and the airport is ten miles, and the distance between the conference center and my hotel is also ten miles. What is the probability that the hotel is closer to the airport than to the conference center?

21. A cube measuring 4 inches on a side is painted. It is then cut into 64 one-inch cubes. One of the smaller cubes is chosen at random and tossed. Find the probability that none of the five faces that are showing is painted.

22. If fifteen distinct whole numbers are randomly selected from 1 to 100, inclusive, determine the probability that two pairs of these numbers will have the same difference.

23. Triangles with sides (a, b, c) are randomly generated in the following manner: $c = 1$, $0 < a \leq 1$, $0 < b \leq 1$. Any value of (a, b, c) that does not satisfy the triangle inequality theorem, $a + b > c$, is discarded. What is the probability (to the nearest hundredth) that a random triangle is obtuse?

24. A box contains fewer than twenty marbles. If you reach into the box and randomly pull out two marbles without replacing them, you have a 50 percent chance of getting two blue marbles. How many blue marbles are in the box?

Geometry

1. Three congruent rectangles are placed to form a larger rectangle as shown, with an area of 1350 cm². Find the area of a square that has the same perimeter as that of the larger rectangle.

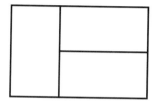

2. How many degrees are in the acute angle formed by the hands of a clock at 2:20 P.M.?

3. The total number of interior angles in two regular polygons is 17, and the total number of diagonals is 53. How many sides does each regular polygon have?

4. What is the maximum number of acute angles that any convex polygon can have?

5. A rhombus has diagonals of lengths 12 and 16 units. What is the perimeter of the rhombus?

6. In a cube with sides of length 1 meter, denote one vertex by the letter A. Find the sum of the distances from A to each of the other vertices of the cube.

7. Ebru's backyard is a square with a side length of twenty meters. In her backyard is a circular garden that extends to each side of her yard. In the center of the garden is a square patch of asparagus so big that each corner of the square touches a side of the garden. Ebru really likes asparagus! How much area in Ebru's garden is not being used to grow asparagus?

8. Find the volume of this figure. All angles that appear to be right angles are right angles.

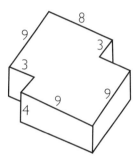

9. If you draw square $ABCD$ and extend the sides by equal lengths to form square $EFGH$, what must be the measure of angle HEB so that the area of $ABCD$ is half the area of $EFGH$?

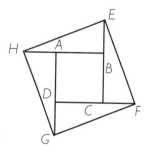

10. *MATH* is a parallelogram in which *SK* = 12, *MK* = (1/3) *MH*, and *ST* = (1/3) *AT*. If the perimeter of *MASK* is 40, find the perimeter of *MATH*.

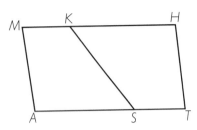

11. A square of maximum area is inscribed in a semicircle as shown. What percent (rounded to the nearest tenth) of the area of the circle is outside the square?

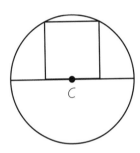

12. A rectangular box measures $6 \times 6 \times 3$ meters. Ashley wants to place an 8-meter pole inside the box. Can she do it?

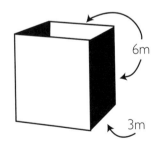

13. A sheet of paper is yellow on the top side and blue on the bottom side. The lower-right corner is folded to the left side of the paper so that the resulting figure is a yellow rectangle above a blue triangle. The width of the paper is 10 cm, and equal areas of blue and yellow appear after the folding. What is the area of one side of the paper?

14. Suppose that you have a square piece of paper on which you draw the largest possible circle. You cut out the circle and discard the leftover scraps of paper. Inside the circle you draw the largest possible square, cut it out, and discard the leftover scraps of paper. How much of the area of the original square remains?

15. Figure *ABCDEF* is a regular hexagon. Points *G*, *H*, *I*, and *J* are midpoints of *FA*, *BC*, *CD*, and *EF*, respectively. Find the ratio of the area of *GHIJ* to that of *ABCDEF*.

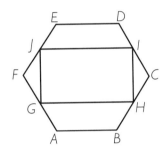

16. Given a sphere inscribed in a cube, find the ratio of the sphere's volume to that of a second sphere circumscribed about the same cube.

17. Line segment *AB* is the diameter of a semicircle. Point *C* lies on the semicircle in such a way that the area of △*ABC* equals half the area of the semicircle. Show that ∠*ABC* measures

$$\frac{1}{2}\sin^{-1}\left(\frac{\pi}{4}\right)$$

radians.

18. The combined volumes of two cubes with integer side lengths are numerically equal to the combined lengths of their edges. What are the dimensions of the cubes?

19. A circle with center C has a diameter of 20 m and a chord RS of length 16 m. Find the length of segment BM.

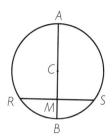

20. The radius of the quadrant and the diameter of the large semicircle are 2. Find the radius of the small semicircle.

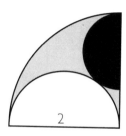

21. What is the area of the union of two circles of radius 1 whose centers are 1 unit apart?

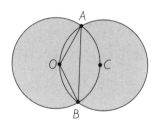

22. The radius of circle O is 3, and the shaded area equals the area of $\triangle ABC$. To the nearest degree, what is the measure of $\angle A$?

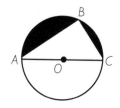

23. In $\triangle PQR$, line STU is parallel to segment PQ. S is on segment RP and T is on segment RQ. Ray PT bisects $\angle RTU$. $ST = 6$ and $PQ = 10$. Find the length of RT.

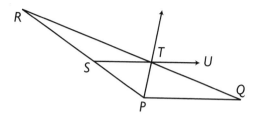

24. The lengths of the sides of a cyclic quadrilateral are 1, 9, 9, and 6, as shown in the diagram. Find $\cos B$.

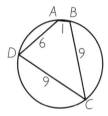

25. Saskatoon (C), Regina (B), and Swift Current (A) form an equilateral triangle. At what point P in the triangle should we build a regional medical facility to minimize the total length of new roads needed to connect P to the existing roads that form the sides of the triangle? The new roads are to be perpendicular to the three existing roads along the sides of the triangle.

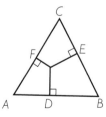

26. A sphere is inscribed in a cone with a base radius of 5 cm and an altitude of 12 cm. Find the radius of the sphere.

27. A volume of oil is put into a spherical bowl of radius 15 cm. After settling, the oil forms a layer that is 5 cm deep. Find the volume of oil.

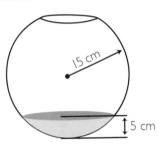

28. Point P is placed inside a rectangle as shown. Show that $a^2 + b^2 = c^2 + d^2$.

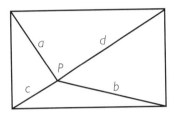

29. If P is moved outside the rectangle, does $a^2 + b^2 = c^2 + d^2$?

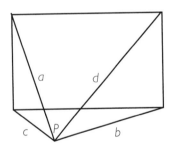

30. P is a point inside quadrilateral $ABCD$. If $PA = 2$ cm, $PB = 3$ cm, $PC = 5$ cm, and $PD = 6$ cm, what is the largest possible area of the quadrilateral $ABCD$?

31. Perpendicular tangents CA and CB are drawn to circle O with radius 4. If $BE = 1$, find AD.

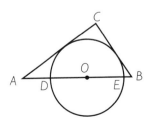

32. Two congruent circles are to be cut from a $9'' \times 12''$ sheet of construction paper. What is the maximum possible radius, to the nearest hundredth of an inch, of these circles? What percent, to the nearest tenth, of the area of the sheet will be cut?

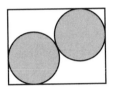

33. This figure is a perspective view of a rectangular solid. $AB = 15$ and $AD = 10$. Find AE so that the number of square units in the surface area of the solid equals the number of cubic units in the volume of the solid.

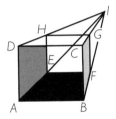

34. Without using Heron's formula, find the area of this triangle.

35. In $\triangle ABC$, BD is the bisector of $\angle B$. Prove that $ac = t^2 + xy$.

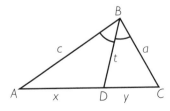

36. Two noncongruent triangles have the same area. One has sides of lengths 5, 5, and 4. The other has sides of lengths 5, 5, and x. Find x.

37. A handcart is used to load a truck. The ramp is 10 feet long and meets the truck 2 feet above the ground. When the cart is tilted backward, the back makes an angle of 80 degrees relative to the ramp. What is the angle, to the nearest tenth of a degree, of the foot of the cart relative to the ground?

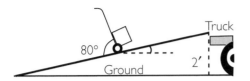

38. Right triangle ABC has legs with lengths 19 and 95 units. The triangle is to be rotated in space about one of its three sides. What is the maximum possible volume of the resulting solid?

39. The T puzzle is made up of four pieces, as illustrated. What is the area of the piece marked x? Note that B and E are the midpoints of segments AC and FD, respectively.

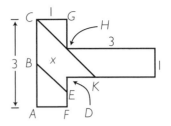

Logs and Exponents

1. Without using a calculator, list the numbers 2^{100}, 3^{75}, and 5^{50} in order from the smallest to the largest.

2. What is the value of x if $4^{20} + 4^{20} = 2^x$?

3. If $n = 3^x + 3^x + 3^x$, then n^2 equals which of the following?

   (a) 9^{3x} (b) 27^{2x} (c) 9^{x+1}

   (d) 27^{6x} (e) 27^{3x}

4. Find the sum:

   $\log_2 8 + \log_4 8 + \log_8 8 + \log_{16} 8 + \log_{32} 8$.

5. Which is larger,

   $$\left(\frac{1}{3}\right)^{1000} + \left(\frac{2}{3}\right)^{1000}$$

   or

   $$\left(\frac{1}{\sqrt{2}}\right)^{1000} + \left(1 - \frac{1}{\sqrt{2}}\right)^{1000} \ ?$$

6. The distance between Earth and the moon is about 400,000 km. Take a piece of cardboard that is 1 mm thick. How many times would you have to fold the same piece of cardboard over and over again, doubling the thickness with each fold, to make it reach a thickness equal to this distance?

7. Solve for x:

   $$\left(x^2 - 5x + 5\right)^{x^2 - 9x + 20} = 1$$

8. If $2^x = 15$ and $15^y = 32$, find the value of xy.

9. Solve for x:

   $$8^{3x+1} - 8^{3x} = 448$$

10. Find the product of all real solutions of

    $$16^{x^2+x+4} = 32^{x^2 + 2x}.$$

11. Compute

    $$\left(\sqrt[3]{\sqrt{30} + \sqrt{3}}\right) \cdot \left(\sqrt[3]{\sqrt{30} - \sqrt{3}}\right).$$

12. Find k if

    $$\sqrt{\frac{a}{b}\sqrt{\frac{b}{a}\sqrt{\frac{a}{b}}}} = \left(\frac{a}{b}\right)^k.$$

13. If $\log_b (xy) = 11$ and $\log_b (x/y) = 5$, what is $\log_b x$?

14. Solve for x:

    $$\sqrt{1 + \sqrt{3 - \sqrt{1 + \sqrt{2 + \sqrt{x}}}}} = 1.$$

15. Find a and b if $(2^a)(9^b)$ equals the four-digit number $2a9b$.

16. Solve the following system:

    $$x^{1/4} + y^{1/5} = 5$$
    $$x^{3/4} + y^{3/5} = 35$$

17. If $P = 2^{1988} + 2^{-1988}$ and if $Q = 2^{1988} - 2^{-1988}$, compute $P^2 - Q^2$.

18. If $\log 2 = M$, $\log 5 = N$, and $\log 7 = P$, express $\log (78.4)$ in terms of M, N, and P.

19. The following is a partial, rounded table of logs to some integer in base a. Complete the table up to $\log_a 10$ and then approximate a.
    Hint: What is $\log_a (c \cdot d)$?

n	$\log_a$
2	0.387
3	0.613
5	0.898
7	1.086

20. Let six numbers in a list be the solutions of $\log_a 81 = 4$, $\log_4 8 = b$, $\log_c (1/27) = -3$, $\log_5 d = 2$, $\log_4 16 = e$, and $\log_9 f = 1/2$. If two values from the list are chosen at random, what is the probability of the occurrence that they are equal?

21. The first two terms of an arithmetic progression are $\log_2 3$ and $\log_2 9$. If the sixth term is x, compute the numerical value of 2^x.

22. How many digits occur in 5^{5^5}?

23. Did you know that all positive integers N can be expressed using three 2s and mathematical symbols? Show that

$$N = -\log_2 \log_2 \sqrt{\sqrt{\ldots \sqrt{\sqrt{2}}}}$$

where N equals the number of root symbols.

Sequences and Patterns

1. If this lattice is continued, what number will be directly below 100?

```
              1
          2   3   4
      5   6   7   8   9
  10  11  12  13  14  15  16
```

2. Let the natural numbers be listed as one continuous sequence of digits without spaces—that is,

$$\{123456789101112\ldots\}.$$

At what locations will the digits 6 and 3 first appear in the order 63?

3. The positive integers are written in an array according to the pattern shown:

Row 1: 1
Row 2: 2   3
Row 3: 4   5   6   7
Row 4: 8   9   10 11 12 13 14 15

The first number in a row equals the number of entries in that row. Determine the row and column in which you would find the number 1996.

4. Consider this tiling pattern. Find an expression for the number of tiles that would be expected in the nth figure. The first three figures are shown.

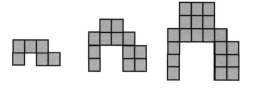

5. How many black tiles will be required to build the twentieth figure in this pattern?

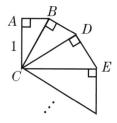

6. Complete this square of numbers:

1	2	3	4
2	5	10	17
3	10	25	52
4	17	52	?

7. In $\triangle ABC$, AC equals 1 and $m\angle ABC$ equals 60 degrees. If this process is continued using 30°-60°-90° triangles, what will be the length of the hypotenuse of the tenth triangle?

8. An n-sided polygon has the property that the measures of its angles in degrees form an arithmetic sequence when listed from smallest to largest. If the smallest angle measures 20 degrees and the largest angle measures 160 degrees, find n.

9. How many arithmetic progressions exist that satisfy all the following conditions:

    1. The progression has at least three terms.

    2. The terms of the progression are all positive integers.

    3. The first term of the progression is 3.

    4. The largest term of the progression is 21.

10. In the sequence of numbers 1, 3, 2, … each term after the first two is defined to be equal to the term preceding it minus the term preceding that. Find the sum of the first one hundred terms of the sequence.

11. A child on a pogo stick jumps 1 foot on the first jump, 2 feet on the second jump, 4 feet on the third jump, …, 2^{n-1} feet on the nth jump. Can the child get back to the starting point by a judicious choice of directions? The measures refer to distances along the ground.

12. Let $N = 1234567891011\ldots998999$ be the natural number formed by writing the integers 1, 2, 3, …, 999 in order. What is the 1996th digit from the left?

13. Twelve positive integers are written in a row. The fourth number is 4 and the twelfth is 12. The sum of any three neighboring numbers is 333. Determine the twelve integers.

14. Find the values of A, B, and C in the four terms of the following arithmetic sequence:

    $AB4, B03, B3C, BA1$.

15. Given that all members of the set of distinct lines $ax + by = c$, with a, b, and c in arithmetic progression, are concurrent at P, find the coordinates of P.

16. What is the following sum?

$$i^{0!} + i^{1!} + i^{2!} + i^{3!} + \ldots + i^{100!}$$

17. The largest square in the diagram has a side of 2. A circle is inscribed in that square, another square is inscribed in that circle, and so on indefinitely. What is the total area of all shaded regions?

18. Evaluate the following expression:

$$\sin^2 1° + \sin^2 2° + \sin^2 3° + \ldots + \sin^2 89° + \sin^2 90°.$$

19. The tenth term of an increasing arithmetic sequence consisting entirely of positive integers is 64. How many distinct sequences possess this characteristic?

20. A square with integral side length s is cut into unit squares, and n of them are shaded to form an X as shown. Write an expression for the unshaded area in terms of s.

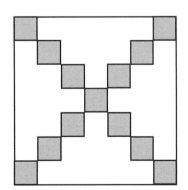

21. If the terms in successive rows of Pascal's triangle are written successively, they form this sequence:

$$\{1, 1, 1, 1, 2, 1, 1, 3, 3, 1, 1, 4, 6, 4, 1, \ldots\}$$

If the sum of the first 212 terms of this sequence is $2^k + k$, find the value of k.

22. In a ten-team baseball league, each team plays each of the others eighteen times. During the season, no game ends in a tie, and at the end of the season, no two teams have the same record. In fact, when the teams are ranked in the order in which they finished, the difference in the number of wins of consecutive teams is a constant. What is the greatest number of games that the last-place team could have won?

23. Patrick, a retired math teacher, loves the taste of coffee in his hot chocolate. He drinks half of his cup of hot chocolate, then fills the cup with coffee. After stirring and drinking another half-cup of the mixture, he again fills the cup with coffee. He continues in this way until he has consumed three cups of the liquid. How much of the original hot chocolate remains in the cup?

24. Find the sum of this series:

$$1(1!) + 2(2!) + 3(3!) + \ldots + n(n!).$$

25. If the integer k is added to each of the numbers 36, 300, and 596, the results are the squares of three consecutive terms of an arithmetic sequence. Find k.

26. What is the sum of all the digits needed to write each counting number from 0 through 1,000,000?

Counting

1. A purse contains 1 quarter, 1 dime, 2 nickels, and 2 pennies. How many different sums of money can be made using at least one of these coins?

2. Myrtle has a small cube. Each face is a different color. She wants to make the cube into a die by putting 1, 2, 3, 4, 5, and 6 on the faces, with the condition that 1 and 6, 2 and 5, and 3 and 4 must be on opposite faces. In how many different ways can she mark the cube?

3. We are given 87 tibbs. All 34 gibbs and all 49 pibbs are tibbs. If exactly 9 tibbs are gibbs and pibbs, then how many tibbs are neither pibbs nor gibbs?

4. Angle *AED* is an acute angle. How many other acute angles are in the diagram?

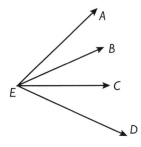

5. Near the end of a party, everyone shakes hands with everybody else. A straggler arrives and shakes hands with only those people whom the straggler knows. Altogether, sixty-eight handshakes occurred. How many people at the party did the straggler know?

6. A graphic artist is creating a logo to represent a company's image. The artist wants to construct a circle that is tangent to all three circles shown. How many different logos can be made? Sorry, the three original circles cannot be arranged differently.

7. How many four-digit numbers contain the digit pattern 75 at least once?

8. We interviewed forty-eight students about recycling paper (P), bottles (B), and cans (C). The following chart shows the number of students who do *not* recycle one or a combination of these items. How many students recycle all three items?

P	B	C	PB	PC	BC	PBC
13	6	9	3	7	4	2

9. A digital clock displays the hour and minute, such as 2:56. How many times between 12:00 midnight and 11:59 A.M., inclusive, will the clock display the numeral 1 at least once?

10. Consider the set of four-digit numbers in which all the digits are distinct. This set includes such numbers as 1492 and 1234 but not such numbers as 1231. If these numbers are listed in order, how many numbers will come before 1239?

11. Given twenty couples, how many different three-member committees can be formed that do not contain any of these couples?

12. Determine the number of integers between 0 and 1000 that contain at least one 2 but no 3.

13. How many different paths lead from point A to point B on the following map if moves only to the right or downward along the sides of the squares are permitted?

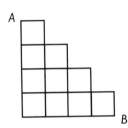

14. How many different sizes of equilateral triangles can be formed on this twenty-five-pin isometric geoboard? What is the area of each triangle if 1 unit of area is represented by the given equilateral triangle?

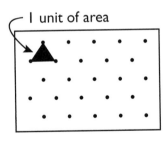

15. In how many ways can six people be assigned to offices if two people refuse to share the same office and if two double offices accommodating two people each and two single offices are available?

16. Different triangles can be made by choosing three vertices from the points A, B, C, D, E, F, and G in the arrangement shown—for example, $\triangle ABF$ and $\triangle ADG$. How many triangles can be made?

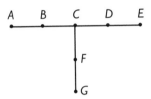

17. Given the nine points shown, determine the number of triangles that can be formed with vertices at the points.

18. Find the total number of distinct rectangles in this figure composed of congruent squares.

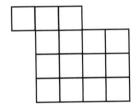

19. How many multiples of 4 from 1 through 10,000 do not contain any of the digits 6, 7, 8, 9, or 0?

20. A square region is divided into eight equal parts, as shown.

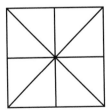

If repeated patterns resulting from flips or spins are not counted, in how many different ways can half the triangular regions be colored with a single color?

21. With the digits 3, 4, 5, 5, 5, 6, and 6, how many distinct integers greater than 5,000,000 can be formed?

22. How many decimal numerals are made up of the digits 1, 2, 3, 4, and 5, each used at most once, which are also multiples of 8?

23. Using the given array, determine the number of ways to spell HANNAH. Always pass from one letter to an adjacent letter. Diagonal, forward, and backward steps are allowed.

H H H H H H
H A A A A H
H A N N A H
H A N N A H
H A A A A H
H H H H H H

Word Problems

1. Allan buys four new tires and a new spare tire for his car. He rotates his tires so that after driving 5000 miles, every tire has been used for the same number of miles. For how many miles was each tire used?

2. A twenty-five-question mathematics test is scored by allowing 5 points for each correct answer and –4 points for each wrong answer. A student loses 3 points if the answer is missing. If a student scored a total of 64 points, how many answers were correct? Wrong? Missing?

3. The square root of half the number of bees in a swarm have flown out to a jasmine bush. In addition, one female bee is flying around a male that got stuck in a lotus flower the evening before. Eight-ninths of the whole swarm remain in the hive. How many bees are involved in activities outside the hive in all?

4. Tom can beat Dick by one-tenth of a mile in a five-mile race. Dick can beat Harry by one-fifth of a mile in a five-mile race. By how much can Tom beat Harry in a five-mile race?

5. Two jars of the same brand of crunchy peanut butter stand on a shelf in a supermarket. The taller jar is twice the height of the other jar but its diameter is one-half as great as the diameter of the shorter jar. The taller jar costs $1.00, and the shorter jar costs $1.50. Which jar is the better buy?

6. An airline allows a passenger's luggage to weigh x pounds and charges an additional fee for each excess pound. Mr. and Mrs. Byrd had a combined luggage weight of 105 pounds and were charged $1.00 and $1.50, respectively. A third passenger also had 105 pounds of luggage and was charged $6.50 for the excess weight. How many pounds of luggage are allowed free of charge for each passenger?

7. Vic has at least one son and one daughter and owns a yacht. His age is more than the number of his children, but he is not yet 100 years old. If the product of his age, the number of his children, and the length in feet of his yacht is 32,118, how old is he, how many children does he have, and how long is his yacht?

8. Twenty men did one-fourth of a job in eight days. Then, because of a building dedication, it became necessary to complete the job in the next five days. How many more men were added to the crew of twenty to accomplish this task?

9. Luke and Slim have only one horse. Luke rides for an agreed-on distance and then ties up the horse for Slim, who has been walking. Meanwhile, Luke walks on ahead. They alternate walking and riding. If they walk 4 miles per hour and ride 12 miles per hour, what part of the time is the horse resting?

10. Two cylindrical water tanks stand side by side. One has a radius of 4 meters and contains water to a depth of 12.5 meters. The other has a radius of 3 meters and is empty. Water is pumped from the first tank to the second tank at a rate of 10 cubic meters per minute. How long, to the nearest tenth of a minute, must the pump run before the depth of the water is the same in both tanks?

11. Under a tree 20 chih high and 3 chih in circumference grows an arrowroot vine that winds seven times around the trunk of the tree and just reaches its top. How long is the vine?

12. A baseball team has $20 million for twenty-five players. All salaries must be multiples of $100,000, and the entire budget must be spent. Each player must receive at least $100,000, but fifteen players must each receive at least $200,000. The shortstop has asked for at least $1 million, and the pitcher for at least $1.5 million. Both demands must be met. Under these conditions, what is the maximum possible salary?

13. If Joe gives Pepe 10 walnuts and Mia gives Pepe 15 walnuts, then Pepe has as many as Joe and Mia together. But if instead Joe gives Mia 15 walnuts and Pepe gives her 8 more, then Mia has as many as Joe and Pepe together. What is the smallest number of walnuts that each person could have?

14. Before going to a garage sale, Elise counted the money she was taking along. After an hour, she counted it again and found that she had spent exactly half of it. The number of cents she now had equaled the number of dollars she started with; the number of dollars she now had was half the number of cents she started with. How much did she spend?

15. It took a train 45 seconds to pass through a 1320-foot tunnel. At the same speed, it took 15 seconds for the train to pass the watchman. How long was the train?

16. In miles per hour, what was the speed of the train in problem 15?

17. A running track has straight parallel sides and semicircular ends. Each lane is 1 meter wide. How much of a head start should the runner in the outside lane receive over the runner is the second lane from the outside so that they will each cover the same distance in one lap around the track?

18. An engineer working on the Alaskan pipeline was heard to make the following remarks:

At the time I said I could finish this section in a week, I expected to get two more bulldozers for the job. But I haven't, and they haven't even left me all the bulldozers I had! If they had, I'd have been only a day behind schedule. As it is, they have taken away all my machines but one, and I'll be weeks late.

How many weeks was he behind?

19. A pencil, eraser, and notebook together cost $1.00. A notebook costs more than two pencils, and three pencils cost more than four erasers. If three erasers cost more than a notebook, how much does each cost?

Puzzles and Games

1. Place the digits 1, 2, 3, 4, 5, 6, 7, and 8 in separate boxes so that boxes that share common corners do not contain successive digits.

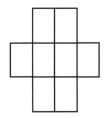

2. The clock shown has a regular hexagonal face. The sum of the numbers along the top side is 24, or 11 + 12 + 1. Rearrange the numbers on the clock so that the sum of the three numbers along any side is 17.

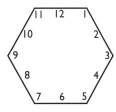

3. Place fifteen dots on the sides of a hexagon so that each side contains the same number of dots.

4. How many triangles are contained in the figure?

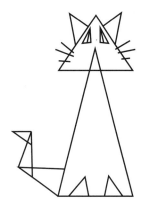

5. $123 - 45 - 67 + 89 = 100$

   This expression equals 100 and uses the digits 1 through 9 in ascending order with the smallest number of plus or minus signs. What is the corresponding expression when using the digits 1 through 9 in descending order?

6. Which of the ten digits does A represent in the following problem? A, S, M, and N are all different.

$$\begin{array}{r} AS \\ \times\quad A \\ \hline MAN \end{array}$$

   Hint: Only the value of A needs to be found.

7. Collect as much money as you can. Enter at the start, exit at the finish, and pass through each room no more than once.

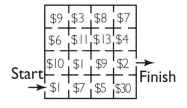

8. $(I) \bullet (AM) \bullet (NOT) = SURE$

   Find at least one solution for a one-digit number times a two-digit number times a three-digit number equaling a four-digit number where all the digits are different.

9. Five coins are placed in a circle, each touching its two neighbors. Two players take turns removing either one coin or two coins that are touching. The winner is the last player to take a coin. Who should win?

10. Remove five of the twelve digits in this addition problem so that the remaining numbers will add to 1111.

$$\begin{array}{r} 111 \\ 333 \\ 777 \\ + \ 999 \\ \hline \end{array}$$

11. The five Olympic rings cut the plane into fifteen pieces not counting the infinite piece on the outside. Arrange the numbers 1 through 15 in these pieces so that the sum of the numbers inside each ring is 39.

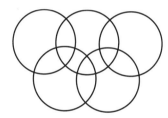

12. A paper with a multiplication example in a non-base-ten system got wet. The only legible numbers remaining are shown. Supply the missing numbers (□), the base of the system used, and the product in base ten.

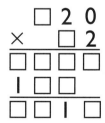

13. Fifteen pennies are placed on a table in front of two players. Each player is required to remove at least one penny but no more than three pennies on his or her turn. The player who removes the last penny loses the game. The players alternate turns until one player takes the last penny on the table and loses all fifteen pennies. Will any method of play guarantee victory? If so, what is it?

14. Place the numbers 2, 3, 4, 5, 6, 7, 8, 9, and 10 in the boxes so that the sum of the numbers in the boxes of each of the four circles is 27.

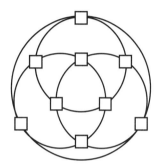

15. Arrange eight identical toothpicks to form a regular octagon, two squares, and eight triangles.

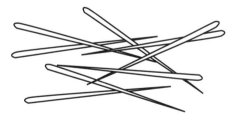

16. Two players take turns removing one, two, or three cards from a pack. A player must never remove the same number of cards as has the previous player. The winner is the one who either takes the last card or leaves the other player with no valid moves. Is it an advantage to start if the pack has four cards? Six cards?

17. Five Rs and five Ws are arranged in a row as shown:

RRRRRWWWWW.

If you can switch just two adjacent letters at a time, find the smallest number of moves that will give you the arrangement

RWRWRWRWRW.

18. Show that when two out of these three coins are turned over at the same time, all heads will never occur, regardless of the number of times two of the three are turned over.

19. A pair of dice features one die with three 0's and three 1's on its faces and another with the numbers 1, 3, 5, 7, 9, and 11 on its faces. This pair has the unusual property that only the sums 1, 2, 3, …, 12 are possible and are equally likely to occur. Design another such pair of six-faced dice that share this property.

20. The ends of three strings are tied to *AC*. Add three more strings. Put one end of each new string through an eye on *BD*; tie the other end to a free end of an original string so that the tangle will become three parallel strings.

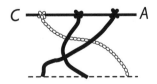

21. Two queens are to be placed on a chessboard of reduced size so that a minimum number of squares are left "unattacked." What is the minimum number of unattacked squares if the board has the dimensions of (*a*) 5×5? (*b*) 6×6?

22. Show how to measure exactly 3 inches using only a sheet of paper that measures 8 1/2 inches by 11 inches.

23. Each letter *P* stands for a prime digit (2, 3, 5, 7). Make the substitutions so that the multiplication is correct.

$$\begin{array}{r} PPP \\ \times\ \ PP \\ \hline PPPP \\ PPPP \\ \hline PPPPP \end{array}$$

24. Ray has 16 mice. Their pen is a 4×4 array with 9 short interior walls creating areas of 8, 3, 3, and 2 square units. He must change the pens to have areas of 6, 6, and 4 square units. He has found a way to rebuild by moving exactly 2 inner walls to different positions or by moving exactly 3, 4, 5, 6, or 7 inner walls. Can you?

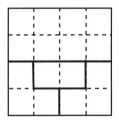

25. Fill a square of integer side *N* with as few smaller squares of integer side lengths (squarelets) as possible. The figure shows a 5×5 example that has been divided into eight squarelets. Into how few squarelets can a 13×13 square be divided?

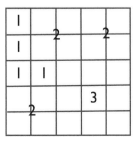

Facts

1. Aristotle used deductive logic—in particular, syllogism. Solve for G to find in which century B.C. he was born.

   $A = D$, $B = 4$, $E = 3$, $D = E$, and $G = A + 1$.

2. Mary Fairfax Somerville was a mathematician in England in the nineteenth century. To discover the year in which she was awarded a silver medal for her solution to the Diophantine equation, find the value of the following expression:

   $$3(1 + 3 + 5 + \ldots + 51) - 221.$$

3. Emilie de Breteuil, French mathematician and friend to Voltaire, published her book *Institutiones de Physique* in the year determined by the following value:

   $$(2^0 + 2^1 + 2^2 + 2^3 + \ldots + 2^9) + 717.$$

4. Augustus De Morgan was alive during the nineteenth century and once wrote, "I was x years old in the year x^2." What was the year of his birth?

5. The Pythagoreans were fascinated by *perfect, deficient, abundant*, and *amicable* numbers. A counting number is said to be perfect if it is equal to the sum of its proper factors. Show that 8128 is a perfect number.

6. The following binary code was created by Francis Bacon. What would the letter Z look like in this code? How many permutations are possible?

   A = aaaaa, B = aaaab, C = aaaba,
   D = aaabb, E = aabaa, F = aabab,
   G = aabba, H = aabbb, I = abaaa, …

7. Joseph Liouville's candidate for a transcendental number was

   a = .110 001 000 000 000 000 000 001….

   Where is the next 1?

8. Gabrielle Émilie du Châtelet is well known for her translation and analysis of Newton's masterpiece, *Principia*. The year of her birth is a four-digit number in which the first two digits are the seventh prime number, the third digit is the additive identity, and the last is the second composite number. du Châtelet lived the same number of years as the fourteenth prime number. Find the year of her death.

9. In the ancient Mayan culture, time was counted simultaneously by a 365-day solar year and a 260-day ritual calendar. Suppose that a year begins with day 1 on both calendars. How many solar years will pass before this situation occurs again? How many ritual years?

10. Z. Morón was the first person to dissect a rectangle into unequal squares. His solution appears here. What is the area of the square marked x?

11. Hypatia was born in October in A.D. 370 in Alexandria, Egypt. Some of her work was with conic sections. A right circular cylinder with radius 7 cm contains 1077 cm³ of water. A rock is immersed in the water, and the water level rises 3 cm without any water being lost. What is the volume of the rock to the nearest cubic cm?

12. Leonardo Da Vinci was interested in the geometry of mechanics and such problems as the following:

    The figure shows a 60-pound weight on one end of a lever with the fulcrum in the middle. Mark the point where the 105-pound weight must be placed to balance the lever horizontally. Disregard the weight of the lever.

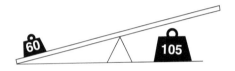

13. Galileo worked with proportional compasses, physics, and astronomy. He might have worked on this problem:

    Two stones are dropped from a building. The mass of stone A is equal to the mass of stone B cubed. If stone A hits the ground with a force of 2048 foot-pounds per second squared, what is the mass of stone B?

14. During the Middle Ages, the square root of a number was estimated by

$$\sqrt{n} = \sqrt{a^2 + b} = a + \frac{b}{2a + 1}.$$

Let $n = 10$. Show why $\sqrt{10}$ was often used as a value for π.

15. In the Sulvasutras ("Rules of the Cord"), a particular number is specified by the following instructions: "To find the number, increase the measure by its third and this third by its own fourth less the thirty-fourth part of that fourth." What number is being described?

16. This problem can be found in the Rhind Papyrus from about 1500 B.C. in Egypt: Divide 100 loaves among five people so that the shares received shall be in arithmetic progression and that 1/7 of the sum of the largest shares shall be equal to the sum of the smallest two. What is the difference of the shares?

17. A tangram is an ancient Chinese puzzle in which a square is cut into seven pieces: five triangles, one square, and one parallelogram. If the area of the entire tangram is 1, what is the area of each piece?

18. Solve this problem from the Michigan papyrus: "Four numbers: their sum is 9900; let the second exceed the first by one-seventh of the first; let the third exceed the sum of the first two by 300; and let the fourth exceed the sum of the first three by 300."

Quickies

1. True or false: An equation of the bisector of the angle formed by the x-axis and the line $y = x$ is

$$y = \frac{1}{2}x.$$

2. Jeremy travels from A to B at 2 minutes per mile and returns over the same route at 2 miles per minute. Find his average speed, in miles per hour, for the whole trip.

3. A number is *abundant* if it is less than the sum of its proper factors. List all two-digit abundant numbers. (There are twenty-one of them.)

4. In what base does 31 equal 2(17)?

5. What is the smallest composite number generated by $p^2 - p - 1$, where p is a prime?

6. If

$$f(n) = \begin{cases} k \text{ when } n = 0 \\ f(n-1) + 20n \text{ when } n > 0, \end{cases}$$

for what value of k will $f(13) = 1996$?

7. Find two positive integers whose product is 24,999,999 and whose (positive) difference is as small as possible.

8. In 1988, the population of Abra increased by 20 percent while the population of Cadabra decreased by 10 percent, after which the two populations were equal. What percent of the original population of Cadabra was the original population of Abra?

9. A group of more than two and fewer than twenty students all paid the same dollar amount for a ticket to a concert. If the total cost was $493, how many students were in the group?

10. The product of a set of distinct positive integers is 48. What is the smallest possible sum of these integers?

11. List all integers less than 20,000 that are both perfect squares and perfect cubes.

12. Given that

$$\frac{23^9 - 23^8}{22} = 23^x,$$

x equals what number?

13. Find three three-digit square numbers that together use each of the digits 1, 2, 3, …, 9 exactly once.

14. Two numbers share an unusual property. They can be expressed as the product of three primes, a, b, and ab, where $a < b$ and ab is a two-digit prime with a as the tens digit and b as the units digit. Find the numbers.

15. At what time between 3 o'clock and 4 o'clock will the big hand be sixteen minutes counterclockwise from the little hand?

16. Triangular numbers can be represented in the following manner:

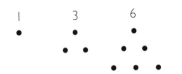

If the difference between a pair of consecutive triangular numbers is 8, what is their product?

17. The area of a rectangle is 360 m². If its length is increased by 10m and its width is decreased by 6m, then its area does not change. Find the perimeter of the original rectangle.

18. You have a large number of stamps in two different denominations. Using combinations of only these two stamps, you cannot pay twenty-seven cents postage but can pay any amount greater than twenty-seven cents. What are the two denominations of the stamps?

19. List these numbers in increasing order:

2^{800} 3^{600} 5^{400} 6^{200}.

20. Each of the 117 crates in a supermarket contains at least 80 and at most 102 apples. What is the smallest number of crates that must contain the same number of apples?

21. Use each of the digits 1 through 9 once to form a fraction equivalent to

(a) $\dfrac{1}{2}$ (b) $\dfrac{1}{3}$.

22. The operation @ is defined as

$$a @ b = a^2 + 3b.$$

Find four pairs of natural numbers such that $a @ b = 37$.

23. Two identical oil pipelines are circular cylinders of radius 6. The Environmental Protection Agency has ordered that they be replaced with a single pipeline with the same total capacity. If the new pipeline is also to be a circular cylinder, what must its radius be?

24. Each of five four-digit primes has the property that the product of its digits is also prime. What are these numbers?

25. How many triangles exist with sides of integer length and a perimeter of 15?

26. A 24-hour digital clock displays the hours and minutes throughout the course of a day. Such times as 2:42 and 13:31 are palindromic. What is the longest time lapse between successive palindromic times?

27. You have $93.97 in the following denominations: two twenty-dollar bills, three ten-dollar bills, four five-dollar bills, one one-dollar bill, one half-dollar, six quarters, five dimes, nine nickels, and two pennies. Can you provide exact change to pay a bill of $46.26 if at least one bill or coin of each available denomination must be used?

28. What is the first two-digit prime number for which the product of its digits is equal to the sum of its digits plus 7?

29. Using only three 9s and any mathematical symbols or operations, write an expression equivalent to 24.

30. If K, E, and D are consecutive integers and if G and B are different consecutive integers, what product is represented by

$$KED = E \times GB?$$

31. If D and E trisect AB, is the area of triangle ABC cut into thirds also? Why, or why not?

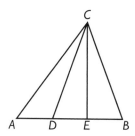

32. Find three different perfect cubes that sum to another perfect cube.

33. The cube of a two-digit number whose digits are different is a five-digit number. This number's digits are all different and also different from those of the original number. Find the five-digit number.

34. Find two dissimilar nonright triangles for which the sides are all integers and the sines of all three angles are rational.

35. Kramer and Erik inherited an Imperial gallon of Grandpa's liquid fish-bait dip. His will instructed how to "divvy it up." Kramer is to pour 1/3 into a container; Erik is to pour 1/3 of the remainder into a second container. They take turns getting 1/3 of the remainder until only 1 insignificant drop remains. Grandpa, a former mathematics teacher, suggested in the will that they find a simpler way to get the same result. Can you?

36. Consider squares $ABCD$, $EFGH$, $IJKL$, $MNOP$, and $QRST$, in which the midpoints of the sides of the squares are the vertices of the inscribed squares. If $AB = 8\sqrt{2}$, find the area of square $QRST$.

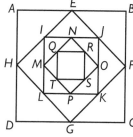

37. If Hypatia randomly takes a coin from her purse, its expected value is 15 cents. If she added a dime to her purse, the expected value of a randomly selected coin would be only 14 cents. What coins does she have in her purse?

38. The top, the front, and one end of a rectangular block have areas of 12 cm², 6 cm², and 8 cm², respectively. What is the volume of the block?

39. Does a Friday the 13th occur every year? How can you be sure?

40. A hexagon is created on a square grid. The vertices are at lattice points. The area is 6 square units. Draw three noncongruent possibilities showing the shape of the hexagon.

41. If $[x]$ denotes the largest integer not exceeding x, where x is any real number, compute the numerical value of

$$\sum_{n=0}^{\infty}\left[\frac{10,000+2^n}{2^{n+1}}\right] = \left[\frac{10,000+1}{2}\right] + \left[\frac{10,000+2}{4}\right] + \left[\frac{10,000+4}{8}\right] + \ldots$$

42. If $a = (2000)(2001)(2002)(2003)(2004)$ and $b = (2002)^5$, which is larger, a or b?

43. If x is 250 percent of y, then what percent of x is $2y$?

44. $\boxed{x}$ = largest prime number less than x.

$\boxed{\boxed{x}}$ = smallest prime number greater than x.

Find the value of the following expression:

$$41 + \boxed{35} - \boxed{\boxed{35}} + \boxed{\boxed{23}}$$

45. Three successive terms in some row of Pascal's triangle,

$$\binom{n}{k-1}, \binom{n}{k}, \text{ and } \binom{n}{k+1},$$

are in a 1:2:3 ratio. Find these three terms.

46. The mean of a set of numbers is 120. If one number is increased by 300, the mean increased to 135. How many numbers are in the set?

47. I had a flat tire after biking 3/4 of the way home. I walked the rest of the way home. If my walking time was twice my biking time, how many times faster is my biking speed than my walking speed?

48. Find all pairs of prime numbers whose sum equals 999.

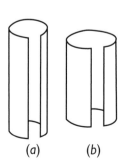

49. A baseless cylinder may be formed in two ways from a sheet of paper that measures 8.5 inches by 11 inches. The two resulting cylinders have the same lateral surface area. Are their volumes also equal?

(a)   (b)

50. Points A, B, C, and D lie on the same line, in that order. If AB:AC = 1:3 and BC:CD = 4:1, compute the ratio AB:CD.

51. The third perfect number is given by the sum of the smallest four consecutive odd integers for which the first is a perfect square; the second, a multiple of three; the third, a perfect cube; and the fourth, a prime. What is the third perfect number?

52. The number 1729 is very interesting: it is the smallest number expressible in two different ways as the sum of two cubes. What are the two ways?

53. A cone has radius x cm and height y cm. A cylinder with radius $2x$ cm has the same volume as the cone. What is the height of the cylinder?

54. How many squares can be made on this geoboard?

55. Triangle ABC is equilateral; D, E, and F are midpoints of its sides; and altitudes CF, AE, and BD intersect at point G. How many right triangles are in the diagram?

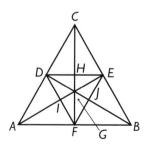

56. Express as a finite or a repeating decimal the number whose binary representation is 1.01010101…, where the dots indicate infinite repetition.

57. Consider all five-digit numbers formed from the five digits 1, 2, 3, 4, and 5, with each digit used at most once in each number. What is the sum of these five-digit numbers?

58. A four-digit number has its digits arranged in the order odd-even-odd-even. If you double the number, all digits are even; if you halve it, only the final digit of the newly formed four-digit number is even. What is the number?

59. Find a two-digit decimal numeral that is twice the product of its digits.

60. An island has no currency; instead, it has the following exchange rate:

50 bananas = 20 coconuts
30 coconuts = 12 fish
100 fish = 1 hammock

How many bananas equal 1 hammock?

61. How many X cards would have to be taken from pile A and put into pile B to make the ratio of X cards to the whole the same for both piles?

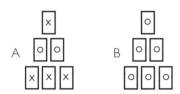

62. Find all digits that have cubes that end with the digit itself.

63. What is the ratio of the length of the tennis-ball can to the circumference of the top of the can? Note that the can holds three balls snugly.

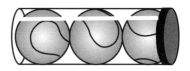

64. A *visible-factor number* is a natural number that is divisible by each of its nonzero digits—for example, 424 or 505. How many visible-factor numbers are less than 100?

65. If the fractions represented by the points R and P are multiplied, what point on the number line best represents their product: M, S, N, P, or T?

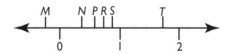

66. Points A, B, C, D, and E lie on a circle. What is the sum of the measures of the interior angles A, B, C, D, and E?

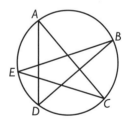

67. A watch shows the calendar date, scrolling through the numbers from 1 to 31 for the days. However, it has to be adjusted to skip over numbers that are not dates in particular months, such as 31 in April. Find the longest stretch that the watch can run without adjustment.

68. Three rectangles are lined up horizontally as shown. The first rectangle has a width of 1 and a length of 2. The second rectangle has a width of 2 and a length of 4. The third rectangle has a width of 4 and a length of 8. Find the area of the shaded region.

69. If P represents the product of all prime numbers less than 1000, what is the value of the units digit of P?

70. How long does it take a clock's hour hand to move through 1 degree of arc?

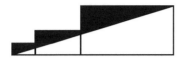

71. Which of these numbers is a perfect square?

329476, 389372, 964328, 326047, 724203

72. One of the rules in the card game cribbage directs players to "score two points for every different combination of cards that totals 15." How many points, for totals of 15, is this hand worth?

(*a*) 16
(*b*) 14
(*c*) 12
(*d*) 10
(*e*) 8

73. When a certain two-digit number is added to the two-digit number having the same digits in reverse order, the sum is a perfect square. Find all such two-digit numbers.

74. What is the smallest sum of the two largest angles in a scalene triangle in which all angle measures are whole numbers?

75. Simplify the following expression:

$$(x - a)(x - b)(x - c)(x - d) \ldots (x - y)(x - z)$$

76. One-quarter is the same part of one-third as one-half is of what number?

77. A 100-square-foot box of plastic wrap costs \$1.29 while a 200-square-foot box costs \$2.19. If each box has an extra 100 square feet added free, which is the better buy?

78. Goldbach's conjecture, not yet proved, states that every even number greater than 2 can be represented as the sum of two prime numbers. Find the smallest possible positive *difference* between two prime numbers whose sum is 98.

79. Both the mean and the median of a set of five distinct natural numbers are 7, and the range is 6. What numbers are included in the set?

80. If $4(9a - 13b) = 6(a - 2b)$, what is the ratio of a to b?

Solutions

SOLUTIONS TO NUMBER THEORY

1. 24.

   Let p represent a prime number greater than 3.

   $$p^2 - 1 = (p - 1)(p + 1)$$

   Since p is prime and greater than 3, either $(p - 1)$ or $(p + 1)$ is divisible by 3. Further, $(p - 1)$ and $(p + 1)$ represent consecutive even numbers. Therefore, either $(p - 1)$ or $(p + 1)$ is divisible by 4. Therefore, $p^2 - 1$ is divisible by $3 \times 2 \times 4 = 24$.

3. Consecutive whole number sums—

   $13 + 14 + 15 + 16 + 17$;
   $10 + 11 + 12 + 13 + 14 + 15$;
   $3 + 4 + 5 + \ldots + 11 + 12$.

   Consecutive integer sums also include—

   $-12 + -11 + -10 + \ldots + 17$;
   $-9 + -8 + -7 + \ldots + 15$;
   $-2 + -1 + 0 + \ldots + 12$;
   $-74 + -73 + -72 + \ldots + 75$;
   $-36 + -35 + -34 + \ldots + 38$;
   $-23 + -22 + -21 + \ldots + 26$.

3. The sum of the proper factors of a prime number, p, is always 1. Since $p > 1$, p must be deficient.

4. 24.

   The numbers 6, 8, 10, 14, 15, and 22 are cute.

5. 89.

   Since $87! + 88! = 87!(1 + 88) = 87!(89)$, 89 is the largest prime in the product.

6. 301.

   The lowest common multiple of 2, 3, 4, 5, and 6 is 60. The number of eggs is 1 more than a multiple of 60 and must be divisible by 7.

7. 765-2387.

   Since $210 = 5 \times 42 = 5 \times 6 \times 7$, the first digits in descending order must be 765. Expressing 336 as a product of prime factors, we get $336 = 2^4 \cdot 3 \cdot 7$. Therefore, the primes in ascending order must be 2, 3, and 7. It follows that $2^3 = 8$ is the composite number in the second-to-last position.

8. 90.

   $2 \times 5 \times 9 = 2^1 \times 5^1 \times 3^2$. Hence, this product has $2 \cdot 2 \cdot 3 = 12$ factors—namely, 2, 5, 9, and nine additional factors, as required.

9. 24.

   A circle has 360 degrees. The lowest common multiple of 75 and 360 is 1800. Therefore, after rotating 1800 degrees, the sector will next land in its original position: 1800 divided by 75 = 24 movements.

10. 2.

    Since
    $$2^3 \equiv 8 \equiv 1 \pmod 7$$
    and
    $$(2^3)^{50} \cdot 2^1 \equiv 1^{50} \cdot 2^1 \equiv 2 \pmod 7,$$
    the remainder is 2.

11. 3.

    By multiplying and keeping track only of the units digit, one can see that a power of 11 must end in the digit 1, a power of 16 must end in the digit 6, and an even power of 14 must end in the digit 6 (odd powers end in the digit 4). Hence, the units digit of the required number is obtained from the sum $1 + 6 + 6$.

12. 30.

    Numbers ending in 75 are divisible by 25. We require that $(a + b)$ is a multiple of 3 if $ab75$ is divisible by 75. Therefore, 1275 is the smallest such number, and repeatedly adding 300 will produce others: 1575, 1875, 2175, …, 9975.

13. 5 and 10; 11 and 14; 37 and 38.

    Trial and error might yield 10 and 5, since $10^2 - 5^2 = 100 - 25 = 75$. To be more systematic, search for natural numbers n and y, $n > y$, such that $n^2 - y^2 = 75$;

    $$\therefore (n - y)(n + y) = 75.$$

    The following table summarizes the results:

$(n - y)$	$(n + y)$	n	y
1	75	38	37
3	25	14	11
5	15	10	5

14. Yes.

    The number 124_{five} is represented by the base-five pieces that follow. Each piece has an odd number of units. The case of 124_{five} has an odd number of pieces, 7, so 124_{five} is odd. In general, if the sum of the digits in base five is odd, the number is odd.

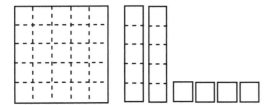

15. $$abba = 1000a + 100b + 10b + a$$
    $$= 1001a + 110b$$
    $$= 11(91a + 10b)$$

16. Fourteen.

    Suppose that the digits are A, B, and C such that $CBA = ABC_R$, where R is greater than or equal to 3, since A, B, and C are unequal.

    $$100C + 10B + A = R^2 \cdot A + R \cdot B + C$$

    $$\therefore \quad 99C = (R^2 - 1)A + (R - 10)B$$

    Values of R from 3 through 13 produce no integral solution (A, B, C) for which A, B, and C are distinct and less than R or 10, whichever is smaller. But $R = 14$ does.

17. 739368 and 739872 are acceptable results. The LCM of 6, 7, 8, and 9 is 504. Since 739000 divided by 504 leaves a remainder of 136, $739000 + (504 - 136) = 739368$ is satisfactory. Adding 504 again, we get 739872.

18. $(2, 1)$, $(9, 3)$, and $(5, 7)$.

    Multiples of 72 must be divisible by both 9 and 8. Hence, the sum of the digits $(1 + 2 + 8 + 7 + x + y + 6)$ must be a multiple of 9 *and* the three-digit number $xy6$ must be a multiple of 8. Therefore, $x + y = 3$ or 12.

    Note that $y = 1, 3, 5, 7,$ or 9 to ensure divisibility by 4. Hence, only the pairs $(x, y) = (2, 1)$, $(0, 3)$, $(9, 3)$, $(7, 5)$, $(5, 7)$, and $(3, 9)$ need to be checked for divisibility by 8. We find that 216, 936, and 576 are multiples of 8.

19. The final two digits of successive powers of 7 beginning with 7^0 are 01, 07, 49, 43, 01, 07.... Since $1997 \equiv 1 \pmod 4$, its final digits will be the same as those of 7^1, or 07.

20. 101 and 9901.

    $$1,000,001 = 100^3 + 1^3$$
    $$= (100+1)(100^2 - 100 + 1)$$
    $$= 101(10,000 - 99)$$
    $$= 101(9901)$$

21. $x + (x + 1) + (x + 2) = 3x + 3 = 3(x + 1)$

    $x + (x + 1) + (x + 2) + (x + 3) + (x + 4)$
    $$= 5x + 10 = 5(x + 2)$$

    $x + (x + 1) + (x + 2) + (x + 3) = 4x + 6$
    $$= 4(x + 1) + 2$$

    Hence, the sum of four consecutive integers, is never divisible by 4.

22. Twenty-nine.

    We require that $B^3 + 9B^2 + 9B + 6 = 32,225$. Use a table, solve graphically with a computer program or a graphing calculator, or use synthetic division.

23. 90,300.

    The sum of all elements of Y can be represented as $(3 + 6 + 9 + ... + 600) + (4 + 8 + 12 + ... + 600) - (12 + 24 + 36 + ... + 600)$.

24. 18.

    Every even number is divisible by 2, thus yielding a factor of 2. Every multiple of 4, or 4, 8, 12, ..., yields an additional factor of 2. Every multiple of 8, or 8, 16, 24, ..., provides another; every multiple of 16, another; and so on. In this problem, the total number of factors of 2 equals $10 + 5 + 2 + 1$, or 18.

25. 3,816,547,290.

    Observe that the tenth digit must be 0. It follows that the fifth digit must be 5. Note that the sum of the digits 1 to 9 is 45. Hence, the first nine digits will be divisible by 9 regardless of the order in which they are placed. Clearly, there are no restrictions on the first digit. Further, the second, fourth, sixth, and eighth digits must be some permutation of 2, 4, 6, and 8. The sum of the first three digits must be a multiple of 3. Therefore, the sum of the next three digits is a multiple of 3 to ensure that the divisibility by 6 is satisfied. Finally, the subsequent three digits must again be divisible by 3. The result can be obtained by combining trial and error with the above observations in a reasonable manner. (Adapted from Bolt [1989]).

26. 1664, 1665, 1666, 1668, 1680.

Let the numbers be $a < b < c < d < e$. We Choose $b - a = c - b = 1$ so that $c - a = 2$. Take $d - c = 2$, the least common multiple of 1 and 2. Then $d - b = 3$, and $d - a = 4$. Take $e - d = 12$, the least common multiple of 2, 3, and 4. Then $e - c = 14$, $e - b = 15$, and $e - a = 16$. Take $e = 1680$, the least common multiple of 12, 14, 15, and 16. Then $d = 1668$, $c = 1666$, $b = 1665$, and $a = 1664$. We can verify that these five numbers have the desired properties.

Lataille (1999) notes that all multiples of a satisfactory set of solutions will also be a solution. Further, {900, 912, 915, 918, 920} is another solution. Solutions for three integers are {2, 3, 4} and for four integers are {6, 8, 9, 12} and {8, 9, 10, 12}. Are there any smaller positive integer solutions?

27. 625, 6,250, 62,500, ...

The number must have at least three digits because 625 is the smallest multiple of 25 beginning with 6. Let an arbitrary $(k + 1)$-digit, positive integer be represented in base ten as

$$a_k 10^k + a_{k-1} 10^{k-1} + \ldots + a_1 10^1 + a_0$$

Let

$$n = 6(10^k) + a_{k-1} 10^{k-1} + \ldots + a_1 10 + a_0,$$

where $k \geq 2$, be the desired number. Then n also must equal

$$25(a_{k-1} 10^{k-1} + \ldots + a_1 10 + a_0).$$

Equating and simplifying give

$$24(a_{k-1} 10^{k-1} + \ldots + a_1 10 + a_0) = 6(10^k),$$

or

$$a_{k-1} 10^{k-1} + \ldots + a_1 10 + a_0 = 10^k/4.$$

Adding $6(10^k)$ to both sides must reproduce n on the left. Thus,

$$n = \frac{10^k}{4} + 6\left(10^k\right)$$
$$= \frac{25}{4}\left(10^k\right)$$
$$= 6.25\left(10^k\right)$$
$$= 625\left(10^{k-2}\right),$$

where k is greater than or equal to 2. Thus, all members of the sequence 625, 6,250, 62,500, ... have the desired property.

28. 18.

$$\begin{aligned} n &= 1995r + 75 \\ &= 57(35r) + 75 \\ &= 57(35r) + 57 + 18 \\ &= 57(35r + 1) + 18 \end{aligned}$$

Therefore, the remainder when n is divided by 57 is 18.

29. $A = 1$, $N = 3$, and $E = 2$.

If the numbers are written in the decimal, or base-ten, system, we get

$$(E + 8N + 64N + 512A)$$
$$- (E + 5N + 25N + 125A)$$
$$= E + 7N + 49N + 343A.$$

Simplifying, we find that $44A - 14N = E$. But A, N, and E are less than 5, and E is even. The only solution is given by $A = 1$, $N = 3$, and $E = 2$.

30. $.\overline{10}_{(two)}$.

The infinite series

$$\frac{1}{2} + \frac{1}{8} + \frac{1}{32} + \frac{1}{128} + \ldots$$

has a sum of

$$\frac{\dfrac{1}{2}}{1 - \dfrac{1}{4}} = \frac{2}{3}.$$

Hence, the bicimal representation of 2/3 is given by $.\overline{10}_{(two)}$.

31. 1, 2, 3, 6, and 17.

Dividing yields

$$\frac{5n + 3}{3n - 7} = 1 + \frac{2n + 10}{3n - 7},$$

so

$$\frac{5n + 3}{3n - 7}$$

is an integer if and only if

$$\frac{2n + 10}{3n - 7}$$

is an integer. But $2n + 10 < 3n - 7$ for every $n > 17$. Thus, we need only to check the first seventeen positive integers, giving us solutions of 1, 2, 3, 6, and 17.

SOLUTIONS TO COORDINATE GEOMETRY

1. 12.

   The lattice points include the various combinations of $(\pm 3, \pm 4)$, $(\pm 4, \pm 3)$, $(0, \pm 5)$, and $(\pm 5, 0)$.

2. 10 cm.

   Let the bottom leftmost vertex of the square be $(0, 0)$. Hence, the centers are $(7, 7)$ and $(15, 1)$, respectively. The length of the segment equals $\sqrt{8^2 + 6^2} = 10$.

3. 10/7 square units.

   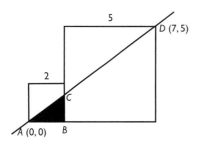

   Equation of AD is $y = 5/7x$. Coordinates of C are $x = 2, y = 10/7$.

   $$\therefore \quad \text{Shaded area} = \frac{1}{2}(2)\left(\frac{10}{7}\right) = \frac{10}{7}.$$

4. $(-3, 0)$, $(1, 8)$, and $(5, 2)$.

5. 36 square units.

   The graph is shown. The area may be calculated in numerous ways.

   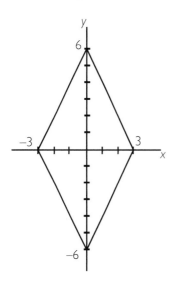

6. $7 + 5\sqrt{2}$.

   If the angles of inclination of $\overset{\longleftrightarrow}{AB}$ and $\overset{\longleftrightarrow}{CB}$ are α and β, respectively, then the angle of inclination of the angle bisector is $(1/2)(\alpha + \beta)$.

   Therefore, the slopes of $\overset{\longleftrightarrow}{AB}$, $\overset{\longleftrightarrow}{CB}$, and the angle bisector are $\tan \alpha$, $\tan \beta$, and $\tan [(1/2)(\alpha + \beta)]$, respectively.

   $$\begin{aligned}
   \text{Slope} &= \tan\left[\frac{1}{2}(\alpha + \beta)\right] \\
   &= \frac{\sin(\alpha + \beta)}{1 + \cos(\alpha + \beta)} \\
   &= \frac{\sin \alpha \cos \beta + \cos \alpha \sin \beta}{1 + \cos \alpha \cos \beta - \sin \alpha \sin \beta} \\
   &= \frac{\dfrac{2}{\sqrt{5}}\left(-\dfrac{1}{\sqrt{10}}\right) + \dfrac{1}{\sqrt{5}}\left(\dfrac{3}{\sqrt{10}}\right)}{1 + \dfrac{1}{\sqrt{5}}\left(-\dfrac{1}{\sqrt{10}}\right) - \left(\dfrac{2}{\sqrt{5}}\right)\left(\dfrac{3}{\sqrt{10}}\right)} \\
   &= \frac{1}{5\sqrt{2} - 7} \\
   &= 7 + 5\sqrt{2}
   \end{aligned}$$

   Alternatively, using a scientific calculator and the following sketch,

   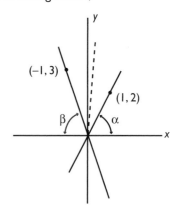

   $$\begin{aligned}
   \alpha &= \tan^{-1} 2 = 63.4349°, \\
   \beta &= \tan^{-1} 3 = 71.5652°, \\
   180° &- (\alpha + \beta) = 45°, \\
   1/2(45°) &= 22.5°, \\
   \alpha + 22.5° &= 85.9349°, \\
   \tan 85.9349° &= 14.0709;
   \end{aligned}$$

   so the slope of the bisector is 14.0709.

7. $y = 1/3x$.

A line that bisects a square must pass through the center of the square—namely, $(3/2, 1/2)$ in this case.

8. $\overline{PA}$ has the area of 5 squares below it, and $\overline{PB}$ has the area of 3 squares below it. Thus, the area of $\triangle PAB$ is equal to the area of 2 squares. We have to choose Q so that $AQ:QB = 1:3$ to get a line with the area of 4.5 squares below it.

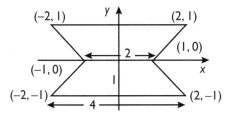

9. $(6, 3)$ and $(2, 1)$.

The analytic conditions are that $(x - 2)^2 + (y - 6)^2 = 25$ and $x = 2y$. Substituting and solving give

$$(2y - 2)^2 + (y - 6)^2 = 25,$$
$$y^2 - 4y + 3 = 0,$$
$$(y - 3)(y - 1) = 0,$$
$$y = 3 \text{ or } 1.$$

The corresponding values of x are 6 and 2.

10. 6 square units.

The interior of the following figure satisfies the conditions. Its area can be calculated in several ways.

(Adapted from Lehigh University High School Mathematics Contest.)

11. $a = -1$ and $b = 2$.

Since $x > -1$, $x + 1 > 0$ and therefore, $|x + 1| = x + 1$. Since $x < 1/2$, $2x - 1 < 0$, and therefore, $|2x - 1| = -2x + 1$. So $y = (x + 1) + (-2x + 1) = -x + 2$. Thus, $a = -1$ and $b = 2$.

12. (d).

The circles begin 14 units apart. During every second, they come 2 units closer to each other, so they meet at 7 seconds. The first circle, centered at the origin, now has a radius of $4 + (7 \times 3)$, or 25; its equation is $x^2 + y^2 = 625$. The second circle, centered at $(30, 0)$, now has a radius of $12 - 7$, or 5; its equation is $(x - 30)^2 + y^2 = 25$. The point $(27, 4)$ satisfies the equation of the second circle.

13. -4.

See the figure for the direct computation of p and q.

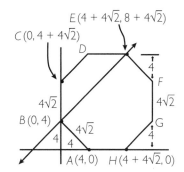

Alternative solution: The line through B and E is given by $y = x + 4$, so (p, q) is on the line if and only if $q = p + 4$. That is, $p - q = -4$.

14. $19/343$.

There are $7 \cdot 7 \cdot 7 = 343$ possible ordered triples (a, b, c). Only those for which a, b, and c are in arithmetic progression will produce an equation $ax + by = c$ with $(-1, 2)$ as a solution. We require that $2b = a + c$. Since c is even, only even values of a are permissible. If $a = -2$ or 2, there are a total of 12 satisfactory triples. If $a = 0$, there are 7 more possibilities, giving a total of 19 satisfactory triples.

1. *Q.*

   To figure out this problem, make a paper cube and arrange the letters exactly the same way you see them in the problem. This way you can find the answer just by looking!

   (From Friedland [1970].)

   The following solution does not use a manipulative. By inspection, note that six distinct letters are represented, so no two sides contain the same letter. We notice from orientations 1 and 2 that if the cube is arranged such that the *S* is vertically aligned, the *A* or the *Q* would be the top face. Thus, the *Q* must be opposite the *A*. To make the puzzle harder, change the *S* to an *O*.

2. Here is one possibility for each:

   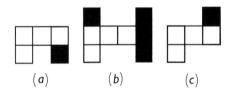

   (a) (b) (c)

3. The distances (in miles) are the real numbers from 2 through 18.

   Without any loss of generality, assume that town A is 8 miles west of town B. Town C could be located any place that is 10 miles from town B (i.e., on a circle, centered at town B, with radius 10 miles). These various positions correspond to distances from town A, which range from 2 miles through 18 miles.

   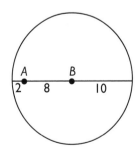

4. (d).

5. (b).

   More than two odd vertices occur in (a), and so the figure cannot be traversed as described. The figure shown in (b) has exactly two odd vertices and can be traversed.

6.

   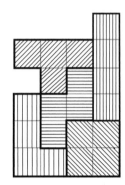

7. (d).
   Try it!

8.

   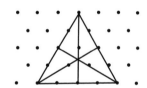

9.

   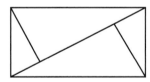

   The four triangles that make up this rectangle are all similar. This division shows one possible solution.

10. *A, D, C, B, H, G, F, E.*

11.

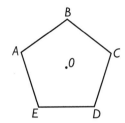

Place 5 points at the vertices of a regular pentagon, and place the sixth point at the center of the pentagon. If the chosen three points all lie on the pentagon, they form the vertices of an isosceles triangle. If the center is chosen, then the other two points must fall on the pentagon. Again they form the vertices of an isosceles triangle.

12. Make a cut from A to C, as shown. The length of this diagonal is $\sqrt{1^2 + 3^2} = \sqrt{10}$.

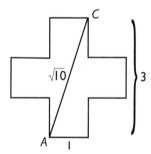

Using one of the pieces, fold vertex A to meet vertex C so that a perpendicular is formed from vertex B to the edge made from the first cut, as shown.

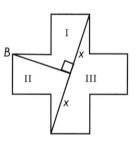

Arrange the three pieces as shown.

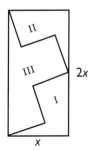

13. 36.

A cube has twelve edges. Each triangular face will add three edges. The total number of edges is 12 + 3(8) = 36.

14. Grids with one even and one odd length can be cut. One dimension in the grid is n, the number of steps in the stairs. If d is the distance to the first step, then the other dimension can be represented by 2d + (n −1). If n is even, 2d + (n − 1) is odd, and vice versa, so that the dimensions of the grid have opposite parity. Those grids that can be cut along grid lines into two congruent stairsteps are only those grids with dimensions of opposite parity.

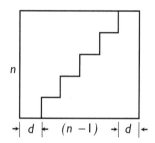

15. One possibility is shown.

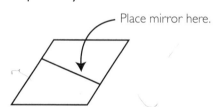

Place mirror here.

16.

Place mirror here.

17.

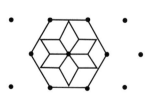

18. Points *A*, *B*, *C*, and *D* in the diagram satisfy both conditions. Points 1 cm from the segment form an "ellipselike" figure, and points 2 cm from the midpoint form a circle.

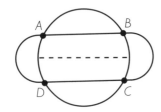

19.

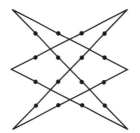

20.

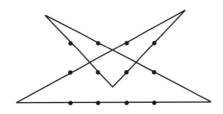

21. No.

To see why not, rotate the 12 × 12 chessboard so that the remaining corner square is in the upper-right corner. Color or shade the board as shown.

The board has forty-eight squares of each color, labeled here as 1, 2, and 3. We have deleted one square of color 1 and two of color 3, so forty-seven of the remaining squares are labeled 1, forty-eight are labeled 2, and forty-six are labeled 3. Thus, the board cannot be covered by 1 × 3 tiles, because for any such tilings, an equal number of squares of each color would be left.

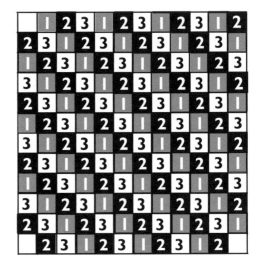

Note that the coloring or shading scheme would need to be revised if the only remaining corner was the lower right. The same general argument can be made.

SOLUTIONS TO LOGIC

1. Five are true; five are false.
   Suppose that statement 1 is true. Then 2 is false; but then 3 is true, and so on. Suppose, however, that statement 1 is false. Then 2 is true; but then 3 is false, and so on. Therefore, we have five of each, but we do not know which is which!

2. Red.
   The key idea is that each box cannot contain a ball of the same color. Therefore, the blue box must have a red ball because the green ball, the only other nonblue ball, is in the red box.

3. Statement 1 is true, statement 2 is false, and statement 3 is true.
   Statement 1 is clearly true. If statement 2 were true, then the other two statements would have to be false. This requirement is not possible, since statement 1 is true. Therefore, statement 2 must be false. This implies that statement 3 is true. Otherwise, it would contradict the fact that statement 2 is false.
   (From Greenes [1986].)

4. Three.
   The last card does not need to be turned over.

5. Actually, no one had the missing nickel. The two methods simply produced different results because equal numbers of pairs and triples were not combined.

6. Black.
   If either of the two lied, then both the man and the woman must have lied.

7. Goblins.
   Suppose the first person was a Spook. Then he or she would have said "Spook" because Spooks tell the truth. Goblins lie and therefore, the first person would have said "Spook" even if he was a Goblin. It follows that the second boy told the truth. Hence, the third was lying, thus making him a Goblin.

8. WKU-UK, OU-OSU, MSU-UCLA, and UT-UNCC.
   Using the fact that if a team is picked by a particular newspaper, its opponent is not picked, we can match UK (the only team on all three lists) with WKU (the only team not on any lists). The only remaining matchup for UT is UNCC. Likewise, OU must meet OSU and UCLA plays MSU.

9. Suppose that no teams played five or more games. Then the maximum number of games in the tournament would equal

   $$\frac{10(4)}{2} = 20.$$

   Since twenty-three games were played, at least one team must have played more than four games.

10. No. It is impossible.
    Each of the eighteen students seated in a black seat must move to a white seat. However, only seventeen white seats are available.

11. The nine nonzero digits form a string that can be neatly divided into three distinct sets of consecutive digits around the circle. These digits sum to 45. Hence, the sets average 45/3 = 15. At least one set of three consecutive digits must sum to at least 15.

12. None.

13. One white ball in one box; one white ball and two black balls in the other. The probability of winning becomes 2/3 with this arrangement.

14. Approach this problem by cutting out graph-paper squares to get the answer. More solutions are possible.

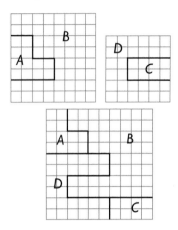

(From Dudeney [1958].)

Therefore,
$$x^2 + y^2 = 2.05xy,$$
$$x^2 + y^2 + 2xy = 4.05xy,$$
$$(x + y)^2 = 4.05xy.$$

ired value

From (1), we get
$$(45)^2 = 4.05xy.$$

Thus,
$$xy = \frac{2025}{4.05} = 500.$$

8. $x = 16, y = 20$.

Let $y - x = k$.

Therefore, $3x + 7y = 10x + 7k = 188$.

∴ $7k$ must end in 8.

∴ $k = 4$ is the smallest possible value.

Letting $k = 4$, we get $x = 16, y = 20$.

9. 1980.

$$5360 = 670\sqrt{280 - 31.6x + x^2}$$

$$8 = \sqrt{280 - 31.6x + x^2}$$

$$64 = 280 - 31.6x + x^2$$

$$x = \frac{31.6 \pm \sqrt{998.56 - 864}}{2}$$

Hence,

$$x = 10 \text{ or } 21.6$$

Since $x = 5$ represents 1975, $x = 10$ represents 1980.

(*Note:* $x = 21.6$ is inadmissible.)

10. 18.
$$K^2 - 3K = -5$$
$$∴ (K^2 - 3K)^2 = 25$$
$$∴ K^4 - 6K^3 + 9K^2 = 25$$
$$∴ K^4 - 6K^3 + 9K^2 - 7 = 18$$

in be

Let the two numbers be x and y.

(1) $x + y = 45$.

(2) $\dfrac{x}{y} + \dfrac{y}{x} = 2.05$.

Equation (2) can be rewritten as

$$\frac{x^2 + y^2}{xy} = 2.05.$$

11. 3.

Let

$$x = \sqrt{6 + \sqrt{6 + \sqrt{6 + \sqrt{6 + \ldots}}}}$$

$$x^2 = 6 + \sqrt{6 + \sqrt{6 + \sqrt{6 + \sqrt{6 + \ldots}}}}$$

$$x^2 = 6 + x$$

$$x^2 - x - 6 = 0$$

$$(x - 3)(x + 2) = 0$$

$$x = 3 \text{ or } -2$$

Since $x > 0$, $x = -2$ is inadmissible; so $x = 3$.

12. $-1/2$.

Substituting $x = k$ and $y = k^2$ into $x + 2y = 0$, we get $k + 2k^2 = 0$. Therefore, $k(1 + 2k) = 0$. The feasible values of k are 0 and $-1/2$. If k equals 0, $(k, k^2, k^3 + 1)$ represents the point $(0, 0, 1)$. This point does not lie on the plane $6x + y + 2z = -1$.

If $k = -1/2$, $(k, k^2, k^3 + 1)$ represents the point $(-1/2, 1/4, 7/8)$. Checking, we find that this point lies on the plane $6x + y + 2z = -1$.

13. $x = -64, y = 0$.

$$x + yi = (1 + i)^{12}$$
$$= \left[(1 + i)^2\right]^6$$
$$= (1 + 2i - 1)^6$$
$$= (2i)^6$$
$$= \left[(2i)^2\right]^3$$
$$= (-4)^3$$
$$= -64.$$

Then

$$x + yi = -64 + 0 \times i.$$

Equating respective real and imaginary parts gives $x = -64$ and $y = 0$.

14. 198.

Let the three-digit number be $abc = 100a + 10b + c$. Then

$$(100a + 10c + b) - (100a + 10b + c) = 9.$$
$$9(c - b) = 9,$$
$$c - b = 1.$$

Also,

$$(100b + 10a + c) - (100a + 10b + c) = 90,$$
$$90(b - a) = 90,$$
$$b - a = 1.$$

The required difference is given by

$$(100c + 10b + a) - (100a + 10b + c) = 99(c - a)$$

But $c - a = (c - b) + (b - a) = 1 + 1 = 2$. Therefore, the number would grow by $99(2) = 198$.

15. 5.

Numerous possible methods of solution are reasonable. You may wish to use physical objects to represent the symbols.

16. 480.

$$(mx + 7)(5x + n) = px^2 + 15x + 14$$
$$5mx^2 + 35x + mnx + 7n = px^2 + 15x + 14$$
$$\therefore 5m = p; (35 + mn) = 15; \text{ and } 7n = 14.$$
Solving, we get
$$m = -10, n = 2, \text{ and } p = -50.$$
$$\therefore m(n + p) = 480.$$

17. 36.

$$6(p - 6) = (p + 6)$$

or

$$6(p - 6) = -(p + 6)$$

Solving, we get

$$p = \frac{42}{5} \text{ or } \frac{30}{7}$$

and

$$\left(\frac{42}{5}\right)\left(\frac{30}{7}\right) = 36.$$

18. 13.5.

$$m^3 + n^3 = (m + n)(m^2 - mn + n^2) = 3(6 - mn)$$

But

$$(m + n)^2 = m^2 + n^2 + 2mn$$
$$\therefore 9 = 6 + 2mn$$
$$\therefore \frac{3}{2} = mn$$
$$\therefore m^3 + n^3 = 3\left(6 - \frac{3}{2}\right) = \frac{27}{2}.$$

19. $c = ab$.

If the sum of two of the roots is 0, then the roots are opposites of each other. Let the roots be s, $-s$, and t. Then

$$x^3 + ax^2 + bx + c = (x - s)(x + s)(x - t)$$
$$= x^3 - tx^2 - s^2x + s^2t.$$

Therefore,

$$c = s^2t = (-b)(-a) = ab.$$

20. In any polynomial $f(x)$, $f(1)$ equals the sum of the coefficients. Hence,

$$f(1) = 1 - 2 + 3 + 4 - 6 = 0.$$

It follows that $(x - 1)$ is a factor of $f(x)$. Therefore, $x = 1$ is a root of the quadratic equation, regardless of how the numbers are assigned to the parentheses.

21. $a = -2$.

We can find a common real solution when $x^2 - ax + 1 = 0$ and $x^2 - x + a = 0$. Subtracting the second equation from the first gives

$$(1 - a)x + (1 - a) = 0,$$
$$(1 - a)(x + 1) = 0.$$

Therefore, $a = 1$ or $x = -1$.

Suppose that $a = 1$. Then $x^2 - x + 1 = 0$, which has nonreal roots, thus eliminating $a = 1$ as a feasible value.

Suppose that $x = -1$. We get $(-1)^2 - a(-1) + 1 = 0$, or $a + 2 = 0$. Therefore, $a = -2$.

Checking, we find that the two equations $x^2 + 2x + 1 = 0$ and $x^2 - x - 2 = 0$ share a common real solution, $x = -1$.

22. When the denominators are equal or the fractions are equivalent.

Let two fractions be a/b and c/d, where b and $d \neq 0$. Let

$$\frac{a+c}{b+d} = \frac{\dfrac{a}{b} + \dfrac{c}{d}}{2}.$$

Therefore,

$$\frac{ad+bc}{2bd} = \frac{a+c}{b+d},$$

$$abd + ad^2 + b^2c + bcd = 2abd + 2bcd,$$
$$ad^2 + b^2c = abd + bcd,$$
$$ad^2 - abd = bcd - b^2c,$$
$$ad(d - b) = bc(d - b).$$

Hence, $ad = bc$ (i.e., $a/b = c/d$) or $d - b = 0$ (i.e., $d = b$).

23. 3.

We have

$$a + bcd = 2$$
$$b + acd = 2$$
$$c + abd = 2$$
$$d + abc = 2.$$

Subtracting the second equation from the first, we find that $(a - b) + (b - a)cd = 0$, or $(a - b)(1 - cd) = 0$. Since $a - b \neq 0$, we must have $cd = 1$. Treating other pairs of equations in the same way (or noting the symmetry in b, c, and d), we find that $bc = 1$ and $bd = 1$. We can then solve the three symmetric equations $bc = 1$, $cd = 1$, and $bd = 1$. Multiplying them, we find $bcd = 1$ or -1. Taking $bcd = 1$ leads to $b = c = d = 1$, and $a = 1$ as well. This outcome contradicts the conditions of the problem. Hence, we must take $bcd = -1$, which leads to $b = c = d = -1$, and $a = 3$.

24. The area of a right triangle with sides of lengths a and b and hypotenuse of length c equals $(1/2)(ab)$. According to Heron's formula,

$$s(s - a)(s - b)(s - c)$$
$$= \left(\frac{a+b+c}{2}\right)\left(\frac{b+c-a}{2}\right)\left(\frac{a+b-c}{2}\right)\left(\frac{a+c-b}{2}\right)$$
$$= \frac{1}{16}(a + b + c)(a + b - c)(c + b - a)(c - b + a)$$
$$= \frac{1}{16}\left[(a + b)^2 - c^2\right]\left[c^2 - (a - b)^2\right]$$
$$= \frac{1}{16}\left(a^2 + b^2 + 2ab - c^2\right)\left(c^2 - (b^2 + a^2 - 2ab)\right).$$

Noting that $a^2 + b^2 = c^2$, we get

$$\sqrt{s(s - a)(s - b)(s - c)} = \sqrt{\frac{1}{16}(2ab)(2ab)}$$
$$= \frac{1}{4}(2ab)$$
$$= \frac{1}{2}(ab).$$

25. Let $(A + B + C + \ldots + I) = 3T$. Therefore, each of the eight columns, rows, or diagonals totals T. Summing all eight of them gives

$$2[(B + H) + (D + F)] +$$
$$3[(A + I) + (C + G)] + 4E = 8T.$$
Since $B + H = D + F = A + I = C + G = T - E$,

$$2(2T - 2E) + 3(2T - 2E) + 4E = 8T$$
$$10T - 6E = 8T$$
$$2T = 6E$$
$$3T = 9E$$

$$\therefore (A + B + C + \ldots + I) = 9E.$$

As required, it has been shown that $E = (A + B + C + \ldots + I)/9$.

26. Three.
$$\frac{26}{65}, \frac{19}{95}, \frac{49}{98}.$$

27. $(2, 1, 1, 2)$.

$$\frac{5}{13} = \cfrac{1}{\cfrac{13}{5}} = \cfrac{1}{2 + \cfrac{3}{5}} = \cfrac{1}{2 + \cfrac{1}{\cfrac{5}{3}}} = \cfrac{1}{2 + \cfrac{1}{1 + \cfrac{2}{3}}} = \cfrac{1}{2 + \cfrac{1}{1 + \cfrac{1}{\cfrac{3}{2}}}} = \cfrac{1}{2 + \cfrac{1}{1 + \cfrac{1}{1 + \cfrac{1}{2}}}}$$

28. 89.

Suppose that a, b, and c represent the number of 40-, 10-, and 1-schilling fruits purchased, respectively. Note that the total purchase price for 100 pieces of fruit is at least 100 schillings in any case. Each 40-schilling fruit costs an extra 39 schillings, and each 10-schilling fruit costs an extra 9 schillings. Thus,

$$259 = 39a + 9b + 100, \text{ or } 159 = 39a + 9b,$$

where a and b are nonnegative integers and $c = 100 - (a + b)$. Then $a = 0, 1, 2, 3,$ or 4, since $39a \leq 159$. Checking, we find that only $a = 2$ produces a feasible solution.

Therefore, $a = 2$ and

$$b = \frac{159 - 39(2)}{9} = 9.$$

It follows that $c = 100 - (2 + 9) = 89$.

29. $(b^2 - 1)/4$.

The quadratic formula gives two values:

$$r_1 = \left(0.5\right)\left(-b + \sqrt{b^2 - 4c}\right)$$

and

$$r_2 = \left(0.5\right)\left(-b - \sqrt{b^2 - 4c}\right).$$

Setting $\left|r_1 - r_2\right| = 1$ and solving yields $c = (b^2 - 1)/4$.

30. (a) The nth term of the sequence is $1/n \times 1/(n + 1)$.
(b) The partial sum is $1/1 \times 1/2 + 1/2 \times 1/3 + 1/3 \times 1/4 + \ldots + 1/n \times 1/(n + 1) = n/(n + 1)$.

31. $x^2 - 6x + 11$.

$$g(x + 3) = x^2 + 2$$
Let $x = a - 3$.
$$\therefore g(a) = (a - 3)^2 + 2$$
$$= a^2 - 6a + 11$$
$$\therefore g(x) = x^2 - 6x + 11.$$

32. $x - 1/3$.

$$f(1 - x) + 2f(x) = x$$
$$\therefore f(1 - x) = x - 2f(x)$$
Replacing x with $(1 - x)$ gives
$$f(x) + 2f(1 - x) = 1 - x$$
or
$$f(x) + 2[x - 2f(x)] = 1 - x$$
$$\therefore -3f(x) = 1 - 3x$$

$$\therefore f(x) = \frac{1 - 3x}{-3} = x - \frac{1}{3}.$$

33. $2/3$, $2/4$, and $2/5$.

Solve $m/n = (m + x)/nx$ to get $m = x/(x - 1)$. Since m must be an integer, $x = 2$ and $m = 2$ or $x = 0$ and $m = 0$. But $m/n > 0$, so the only choice is $m = 2$ and $x = 2$. Next, think about n, which must be such that $1/3 < m/n < 1$, so $1/3 < 2/n < 1$. Thus, $n > 2$ and $n < 6$. Since n is an integer, $n = 3, 4,$ or 5.

34. 643.

We require that

$$x \equiv 3 \pmod 8$$
$$x \equiv 4 \pmod 9$$
$$x \equiv 5 \pmod{11}$$
$$x \equiv 6 \pmod{13}.$$

Observe that 8 and 9 are relatively prime and 3 and 4 are both 5 less than 8 and 9, respectively. Hence, $x \equiv 67 \pmod{72}$ satisfies the first two congruences.

A detailed solution appears in the original calendar. The solution to the congruence is $x \equiv 643 \pmod{10296}$. Note that 8, 9, 11, and 13 are all relatively prime to one another and their product is 10296.

35. $\dfrac{162 + 36\sqrt{14}}{25}.$

We have

$$5x^2 - 12xy - 18y^2 = 0.$$

Dividing by y^2, we get

$$\frac{5x^2}{y^2} - \frac{12x}{y} - 18 = 0.$$

Therefore,

$$5\left(\frac{x}{y}\right)^2 - 12\left(\frac{x}{y}\right) - 18 = 0.$$

Using the quadratic formula, we find that

$$\frac{x}{y} = \frac{6 \pm 3\sqrt{14}}{5}.$$

Therefore, the maximum value of

$$\frac{x^2}{y^2} = \left(\frac{6 + 3\sqrt{14}}{5}\right)^2$$

$$= \frac{162 + 36\sqrt{14}}{25}.$$

SOLUTIONS TO PROBABILITY

1. 0.01.
   The probability is the ratio of the area of a one-meter circle to the area of a ten-meter circle.

2. 5/12.
   The probability equals the sum of the respective probabilities of getting totals of 2, 3, 5, 7, or 11, which is (1 + 2 + 4 + 6 + 2)/36.

3. 2/5, or .4.
   Select a vertex randomly. It is a vertex in

   $$\binom{5}{2} = 10 \text{ triangles,}$$

   four of which are actually isosceles.

4. .000363.
   Of the 9×10^9 possibilities, exactly $9 \times 9!$ contain ten different digits. (Note that 0 cannot be the first digit.)

5. 1/7.
   Seven equally likely outcomes (HTT, THT, TTH, HHT, HTH, THH, HHH) satisfy the given condition. Only HHH is favorable.

6. 0.
   The sum of the digits 1, 9, 9, and 8 is 27. Hence, the four-digit number will be a multiple of both 3 and 9.

7. Six white and four black marbles.
   Note that the conditions tell us indirectly that the chance that the third marble is white would be 1/2 if two white marbles had already been drawn. Hence, the box contains two more white (w) marbles than black (b) marbles. The total number of marbles can be represented as $2w - 2$ because $b = w - 2$.
   Hence,

   $$\frac{\binom{w}{2}}{\binom{2w-2}{2}} = \frac{1}{3}.$$

   This expression simplifies to

   $$\frac{w(w-1)}{(2w-2)(2w-3)} = \frac{1}{3}.$$

Factoring out $(w - 1)$, we get

$$\frac{w}{2(2w-3)} = \frac{1}{3},$$
$$2(2w-3) = 3w,$$
$$\therefore \quad w = 6.$$

Hence, $b = 6 - 2 = 4$.

8. 11/15.
   Observe that each set of thirty positive integers (1–30, 31–60, 61–90, …) will behave similarly. Hence, we can consider the integers 1 to 30, inclusive. The integers N for which GCF (N, 30) > 1 are listed: 2, 3, 4, 5, 6, 8, 9, 10, 12, 14, 15, 16, 18, 20, 21, 22, 24, 25, 26, 27, 28, 30. The required probability is 22/30, or 11/15.

9. 6/25, or 0.24.
   Twenty-five prime numbers, 2, 3, 5, 7, 11, 13, 17, 19, 23, 29, 31, 37, 41, 43, 47, 53, 59, 61, 67, 71, 73, 79, 83, 89, and 97, occur, six of which contain the digit 9.

10. 5/16.
    The probability of three heads twice is $(1/8)^2$, two heads twice is $(3/8)^2$, one head twice is $(3/8)^2$, and no heads twice is $(1/8)^2$. These add to a total probability of 20/64, or 5/16.

11. 35/36.
    There is a 1/36 probability that the same number will appear three times. The desired result is its complement, $1 - 1/36$.

12. 720.
    Each face will appear once in $6 \times 5 \times 4 \times 3 \times 2 \times 1$, or 720 ways. There is only one way of getting six 6's.

13. .40.
    The desired probability equals (.55)(.40) + (.30)(.60)
    (Adapted from Brousseau [n.d.].)

14. 13.
    It is the only sum that can be reached regardless of whether the next to last sum is 8, 9, 10, 11, or 12.

15. 2/21.

$$\binom{9}{3} = \frac{9!}{3!\,6!} = 84$$

Thus, 84 possibilities must be considered. Eight of these possibilities—three columns, three rows, and two diagonals—produce a ticktacktoe win. The probability is 8/84, or 2/21.

16. 4/45.

Ninety three-digit palindromes exist—namely, 101, 202, 303, ..., 909; 111, 212, ..., 919; ...; 191, 292, ..., 999. A palindrome of the form *aba* is a multiple of 11 if $2a - b = 0$ or 11. The numbers satisfying this requirement are 121, 242, 363, 484, 616, 737, 858, and 979. The desired probability equals 8/90, or 4/45.

17. 36/91, 30/91, and 25/91 for players A, B, and C, respectively.

$$P(\text{A wins}) = \frac{1}{6} + \frac{5}{6} \times \frac{5}{6} \times \frac{5}{6} \times \frac{1}{6} + \left(\frac{5}{6}\right)^{6} \times \frac{1}{6} + \dots$$

$$P(\text{B wins}) = \frac{5}{6} \times \frac{1}{6} + \left(\frac{5}{6}\right)^{4} \times \frac{1}{6} + \left(\frac{5}{6}\right)^{7} \times \frac{1}{6} + \dots$$

$$P(\text{C wins}) = \left(\frac{5}{6}\right)^{2} \times \frac{1}{6} + \left(\frac{5}{6}\right)^{5} \times \frac{1}{6} + \left(\frac{5}{6}\right)^{8} \times \frac{1}{6} + \dots$$

These expressions are infinite geometric series with common ratios of $(5/6)^3$.

18. 0.618.

The given condition is that $p^2 = 1 - p$. The positive solution of this quadratic equation is the golden ratio

$$\frac{\sqrt{5} - 1}{2} \approx 0.618.$$

19. 3/10.

We must consider

$$\binom{5}{3} = 10$$

possibilities. The sum of the lengths of the two shorter rods must exceed the length of the longest rod. Only three combinations satisfy this requirement—namely, (15, 30, 40), (30, 40, 60), and (40, 60, 90).

20. 1/3.

Pick a point for the hotel (*H*) and another point ten miles away for the conference center (*C*). The airport (*A*) is on a circle (call it C_1) centered at *C*, with a radius of ten miles. The problem is to find the proportion of the points of C_1 that are closer than ten miles to *H*. That is, what proportion of the points of C_1 are inside a circle (call it H_1) centered at *H*, with a radius of ten miles? The hotel, the conference center, and either of the two intersections of C_1 and H_1 (call the intersections *P* and *Q*) form an equilateral triangle, so $\angle PCQ$ is $2 \times 60° = 120°$. Because 120/360 = 1/3, one-third of the points of C_1 are inside H_1. Therefore, the probability that the hotel is closer to the airport than to the conference center is 1/3.

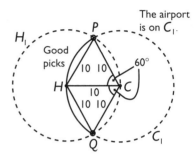

21. 3/16.

Of the 64 one-inch cubes, 8 have no painted faces, and 24 have one painted face. The probability of choosing one with no painted faces is 8/64. The probability of choosing one with one face painted and tossing it so that the painted face does not show is

$$\frac{24}{64} \cdot \frac{1}{6} = \frac{4}{64}.$$

Therefore, the probability that no painted faces will show when a cube chosen at random is tossed is

$$\frac{8}{64} + \frac{4}{64} = \frac{12}{64} = \frac{3}{16}.$$

22. 1.

There are $\binom{15}{2} = 105$ possible pairings of the fifteen numbers. However, there are only 99 possible differences (i.e., 1, 2, 3, ..., 99). Therefore at least two pairs of the numbers have the same difference.

23. Approximately 0.57.

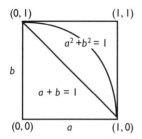

On the coordinate plane, plot (a, b) on (x, y). Points on or to the lower left of the line $a + b = 1$ do not satisfy the triangle inequality theorem and are discarded. If $a^2 + b^2 > 1$, the triangle is acute. Points within the area bounded by $a + b = 1$ and $a^2 + b^2 = 1$ represent obtuse triangles. The ratio of this area to the total area of valid triangles is

$$\frac{\frac{1}{4}(\pi) - \frac{1}{2}}{\frac{1}{2}}$$

or

$$\frac{1}{2}(\pi) - 1 \approx 0.57.$$

Perhaps one might be surprised that more of these randomly generated triangles are obtuse than acute, and the probability is 0 that a right triangle might be generated.

24. Three.

Three blue marbles and one nonblue marble are in the box. Other possibilities exist if the restriction on the total number of marbles (i.e., fewer than 20) is lifted.

Suppose that b blue marbles and a nonblue marbles are in the box. We require that

$$\frac{\binom{b}{2}}{\binom{a+b}{2}} = \frac{1}{2}$$

where $a + b < 20$. Observe that

$$\binom{b}{2}$$

and

$$\binom{a+b}{2}$$

both represent triangular numbers. Therefore, they represent values in the sequence $\{1, 3, 6, 10, 15, \ldots\}$. In fact,

$$\binom{b}{2} = 3$$

and

$$\binom{a+b}{2} = 6$$

are the only values satisfying the requirements. Hence, $b = 3$ and $a = 1$.

SOLUTIONS TO GEOMETRY

1. 1406.25 cm².
The area of each smaller rectangle is 450 cm².
Since the length of these rectangles is twice
the width, we find

$$l \cdot w = 2w^2 = 450 \text{ cm}^2$$
$$\therefore \qquad w = 15 \text{ cm}$$

and
$$l = 30 \text{ cm}.$$

The perimeter of the larger rectangle is
$3l + 4w = 150$ cm. The area of the square is
$(150/4)^2 = 1406.25$ cm².

2. 50.
At 2:20 P.M., the hour hand would be one-third
of the way from 2 to 3. We shall say that it is
at 2 1/3. The minute hand will be at 4. Hence,
the measure, in degrees, of the acute angle is
given by

$$\frac{4 - 2\frac{1}{3}}{12} \times 360 = 50.$$

3. 11 and 6.
If m and n are the numbers of sides of the two
regular polygons, then $m + n = 17$, and

$$\frac{n(n-3)}{2} + \frac{m(m-3)}{2} = 53.$$

Solving, we get $m = 11$ and $n = 6$, or vice
versa.

4. 3.
In a convex polygon, each interior angle is
either acute, right, or obtuse. An interior angle
of measure more than 180 degrees would
contradict the convexity of the polygon.
  In a convex polygon, the sum of the measures of the exterior angles is equal to 360
degrees. The angle exterior to an acute interi-
or angle is an obtuse angle. The sum of the
measures of four or more obtuse angles
exceeds 360 degrees. Therefore, the maximum
number of obtuse exterior angles, or acute
interior angles, is three.

5. 40 units.
The diagonals form four right triangles with
sides of lengths 6, 8, and 10. The perimeter
must be 4×10 units.

6. $(3 + 3\sqrt{2} + \sqrt{3})$ meters.
Of the remaining seven vertices, three form
edges of the cube when joined to A, three
form face diagonals when joined to A, and one
forms a space diagonal of the cube when
joined to A. Hence, the sum of the distances
from A to the other vertices of the cube is
$3(1) + 3(\sqrt{2}) + 1(\sqrt{3})$ meters.

7. $(100\pi - 200)$ sq. meters.

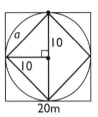

Area of asparagus patch

$$= a^2 = 10^2 + 10^2 = 200.$$

Total area of garden

$$= \pi(10)^2 = 100\pi.$$

$\therefore (100\pi - 200)m^2$ equals the area not used
for asparagus.

8. 492 cubic units.

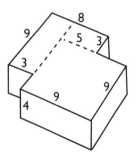

$9 \times 9 + 9 \times 3 + 5 \times 3 = 123$ square units.

Volume = 123 square units $\times$ 4 units
$$= 492 \text{ cubic units.}$$

9. 75 degrees or 15 degrees.
   Label the legs of the triangles $a = EC$ and $b = CF$, as shown, where $a > b$.

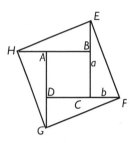

The measures of the sides of the small square are now $(a - b)$. The measures of the sides of the large square are $\sqrt{a^2 + b^2}$. Set up an equation:

$$(a-b)^2 = \frac{1}{2}\left(\sqrt{a^2+b^2}\right)^2$$
$$= \frac{1}{2}\left(a^2+b^2\right)$$

Simplifying, we can rearrange the equation to get $a^2 - 4ab + b^2 = 0$. Applying the quadratic formula gives us an expression for a:

$$a = \frac{4b \pm \sqrt{16b^2 - 4b^2}}{2}$$
$$= 2b \pm b\sqrt{3}$$
$$= b\left(2 \pm \sqrt{3}\right)$$

$$\therefore \frac{a}{b} = 2 + \sqrt{3}, \text{ since } \frac{a}{b} > 1.$$

Thus, arctan $(2 + \sqrt{3}) = m\angle EFC = m\angle HEB = 75$ degrees. Note that the way that a and b have been labeled in this diagram would produce a result of 75 degrees. However, it is possible to draw the diagram so that an angle of 15 degrees produces the required halving of the large square.

10. 56 units.

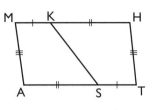

$$MK + 12 + SA + AM = 40$$
$$MK + SA + AM = 28.$$

But $ST = MK$, so $ST + SA + AM = 28$. But $ST + SA = AT$.

Therefore,
$$AT + AM = 28,$$
and
$$2AT + 2AM = 56.$$

11. 74.5 percent.

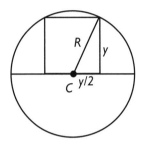

$$R^2 = y^2 + \left(\frac{y}{2}\right)^2$$
$$= \frac{5}{4}y^2,$$
$$\therefore y^2 = \frac{4}{5}R^2,$$

which is the area of the square.

The area of the circle outside the square is

$$\pi R^2 - \frac{4}{5}R^2.$$

The ratio is

$$\frac{\pi R^2 - \frac{4}{5}R^2}{\pi R^2} = 74.5 \text{ percent.}$$

12. Yes.

A 9-meter pole is the maximum length. Apply the Pythagorean theorem twice or use the generalized Pythagorean theorem:

$$\text{diagonal} = \sqrt{6^2 + 6^2 + 3^2} = 9.$$

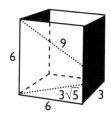

13. 150 cm².

$$\text{Blue area} = 1/2 \cdot 10 \cdot 10 = 50$$
$$\text{Yellow area} = 50 = 10 \cdot x,$$
$$5 = x.$$

Hence, the area of the paper = $10 \times 15 = 150$.

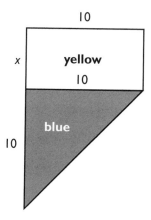

14. One-half the area of the original square remains.

The area of the large square is $4x^2$, the area of the circle is πx^2, and the area of the small square is $(x\sqrt{2})^2 = 2x^2$.

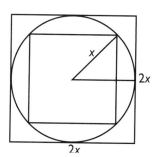

15. 1:2.

Extend *AB, HI, DE,* and *JG* to meet at points *K, L, M,* and *N*. Construct *CF,* intersecting *JG* at *O* and *HI* at *P*.

Since triangles *JOF, JNE, IPC, IMD, GKA, GOF, HLB,* and *HPC* are all congruent, we can see that the area of rectangle *OPIJ* is one-half that of trapezoid *FCDE* and that the area of rectangle *OPHG* is one-half that of trapezoid *FCBA*. Therefore, the area of *GHIJ* is one-half that of hexagon *ABCDEF*.

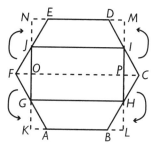

16. 1: 3√3.

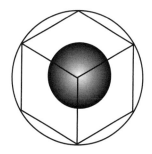

To find the volume of the inscribed sphere, consider that if the length of the edge of the cube is x, then the radius of the sphere is $(1/2)x$. The volume would be

$$V = \frac{4}{3}\pi r^3$$
$$= \frac{4}{3}\pi \frac{1}{8} x^3$$
$$= \frac{1}{6}\pi x^3.$$

To find the volume of the circumscribed sphere, consider that the length of the radius of the sphere would be one-half the main diagonal of the cube. The main diagonal of the cube is given by

$$\sqrt{x^2 + x^2 + x^2} = \sqrt{3x^2} = x\sqrt{3}.$$

The radius is $(1/2)x\sqrt{3}$. The volume is

$$V = \frac{4}{3}\pi\left(\frac{1}{2}x\sqrt{3}\right)^3$$
$$= \frac{1}{6}\pi\left(3\sqrt{3}\right)x^3.$$

Therefore, the ratio of the volume of the inscribed sphere to that of the circumscribed sphere of a cube with edge x is $1 : 3\sqrt{3}$.

17.

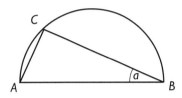

Let the radius of the circle be r and $m\angle B = a$ radians. Then half of the area of the semicircle is $\pi r^2/4$. Since $\angle C$ is a right angle because it is inscribed in a semicircle, $AC = 2r \sin a$ and $BC = 2r \cos a$. Then the area of the triangle is $BC \times AC/2$, or $2r^2 (\cos a) \cdot (\sin a)$. By the double-angle identity, this result is $r^2 \sin (2a)$. Setting the two area expressions equal, we get $\pi r^2/4 = r^2 \sin (2a)$, or $\pi/4 = \sin (2a)$. Hence $a = 1/2 \sin^{-1}(\pi/4)$.

18. 2 and 4 units.
Let x and y be side lengths, where $0 < x \le y$.
$$\therefore x^3 + y^3 = 12(x + y).$$

Dividing by x, we get
$$x^2 - xy + y^2 = 12$$
$$\therefore x^2 - 2xy + y^2 = 12 - xy$$
$$\therefore (x - y)^2 = 12 - xy.$$

Since $0 < x \le y$ and $12 - xy \ge 0$, $0 < x^2 < 12$. Therefore, $x = 1, 2,$ or 3. Checking, we find that $x = 2, y = 4$ is the only integer solution.
(Adapted from Dunn [1980].)

19. 4m.

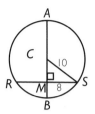

Radius $CS = \frac{1}{2}(20) = 10$,

and

$$MS = \frac{1}{2}(16) = 8.$$
Therefore,
$$CM = \sqrt{10^2 - 8^2} = 6.$$
Hence,
$$BM = 10 - 6 = 4.$$

20. $r = 2/3$.
The segment joining the centers intersects the circles at their point of tangency. By the Pythagorean theorem,

$$(1 + r)^2 = 1^2 + (2 - r)^2.$$

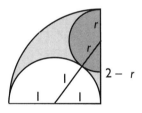

21. $(4\pi/3 + \sqrt{3}/2)$ square units.
Since the radii of the circles are 1 and the centers are 1 unit apart, $\triangle AOC$ and $\triangle BOC$ are equilateral triangles and $m\angle AOB = 120°$. The area of circle O is $\pi \cdot 1^2$. The sector AOB has area

$$\frac{120°}{360°}\pi = \frac{\pi}{3}.$$

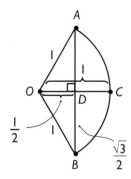

Area $\triangle AOB = \dfrac{1}{2} \cdot AB \cdot OD$

$\qquad = \dfrac{1}{2} \cdot \sqrt{3} \cdot \dfrac{1}{2}$

$\qquad = \dfrac{\sqrt{3}}{4}.$

The area of the union is the area of the two circles minus the two overlapping regions.

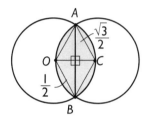

Area $= 2 \cdot \pi \cdot 1^2 -$
$\qquad 2 \cdot$ (area of sector $-$ area of triangle)

$= 2\pi - 2\left(\dfrac{\pi}{3} - \dfrac{\sqrt{3}}{4} \right)$

$= \dfrac{4\pi}{3} + \dfrac{\sqrt{3}}{2}$ square units.

22. **26 degrees or 64 degrees.**

Since $\triangle ABC$ is a right triangle with hypotenuse 6, the legs have lengths 6 cos A and 6 sin A. So

area of $\triangle ABC = \dfrac{1}{2} \cdot 6 \sin A \cdot 6 \cos A$

$\qquad = 9 \cdot 2 \sin A \cos A$

$\qquad = 9 \sin 2A.$

Because the area of $\triangle ABC$ is half the area of the semicircle,

$$9\sin 2A = \dfrac{1}{4} \cdot 9\pi$$

$$\sin 2A = \dfrac{\pi}{4}$$

$$A = \dfrac{1}{2}\sin^{-1}\dfrac{\pi}{4}.$$

Therefore, $m\angle A = 26°$ or $m\angle A = 64°$. Actually, the fact that the radius is 3 is irrelevant. The same result holds no matter what the size of the circle. This problem can also be solved with such geometry software as the Geometer's Sketchpad or Cabri.

23. **15.**

It can be shown that $\angle TPQ$ and $\angle PTQ$ are congruent. Therefore, $TQ = 10$.

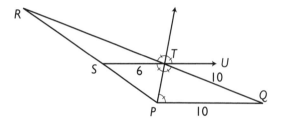

$$\dfrac{ST}{PQ} = \dfrac{RT}{RT + TQ}$$

$$\dfrac{6}{10} = \dfrac{RT}{RT + 10}$$

$$10(RT) = 6(RT) + 60$$

$$4(RT) = 60$$

$$RT = 15$$

24. **−5/18.**

Draw diagonal AC and apply the law of cosines twice to obtain

$$\begin{cases} AC = 1^2 + 9^2 - 2 \cdot 1 \cdot 9 \cos B \\ AC = 6^2 + 9^2 - 2 \cdot 6 \cdot 9 \cos D, \end{cases}$$

or

$$1^2 + 9^2 - 2 \cdot 1 \cdot 9 \cos B$$
$$= 6^2 + 9^2 - 2 \cdot 6 \cdot 9 \cos D.$$

Since $\angle B$ and $\angle D$ are supplementary,
$-\cos B = +\cos D$.

$$-126\cos B = 35$$
$$\cos B = \frac{-35}{126}$$
$$\cos B = \frac{-5}{18}.$$

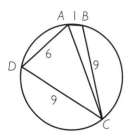

25. Let P be any point in the interior of the equilateral triangle ABC of side s. Draw PD, PE, and PF perpendicular to AB, BC, and CA, respectively. Draw PA, PB, and PC. The area of $\triangle ABC$ = area of $\triangle APB$ + area of $\triangle BPC$ + area of $\triangle CPA$. Thus, if h is the height of $\triangle ABC$, then

$$\frac{1}{2}sh = \frac{1}{2}s(PD) + \frac{1}{2}s(PE) + \frac{1}{2}s(PF),\text{ and}$$
$$h = PD + PE + PF.$$

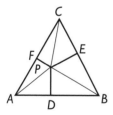

Hence, the sum of these three perpendiculars is constant and equal to h for any point P in the region. The facility can be built at any point in the region, and the total length of new roads to be built to connect the three sides is h.

(Adapted from "About the Cover," *Mathematics Teacher* 85 [April 1992], p. 249; "Cover Activity: Altitude of an Equilateral Triangle," *Mathematics Teacher* 85 [April 1992], p. 250, and Munirv Zabanch, "Reader Reflections: Ideas on April 1992," *Mathematics Teacher* 86 [March 1993], p. 194.)

26. 10/3.

The radius and a portion of the altitude form a right triangle with a segment on the surface of the cone. Therefore,

$$(12 - x)^2 = x^2 + 8^2,$$
$$144 - 24x + x^2 = x^2 + 64,$$
$$24x = 80$$
$$x = \frac{80}{24} = \frac{10}{3}.$$

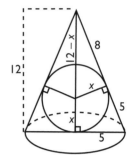

27. $(1000\pi/3)$ cm³.

The volume of a layer in a spherical bowl is $V = (\pi/3)t^2(3r - t)$, where r is the radius of the sphere and t is the depth of the layer. Evaluating

$$\frac{\pi}{3}(5)^2\big[3(15) - 5\big]$$

gives the volume.

28.

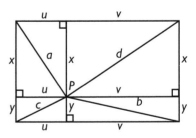

$a^2 + b^2 = (u^2 + x^2) + (v^2 + y^2)$
$c^2 + d^2 = (u^2 + y^2) + (v^2 + x^2)$
$\therefore a^2 + b^2 = c^2 + d^2.$

29. Yes.

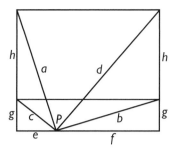

$a^2 + b^2 = [(g + h)^2 + e^2] + (f^2 + g^2)$
$c^2 + d^2 = (g^2 + e^2) + [(g + h)^2 + f^2]$
∴ $a^2 + b^2 = c^2 + d^2$.

30. 31.5 cm².

The area of the quadrilateral is the sum of the areas of triangles PAB, PBC, PCD, and PDA. It is a consequence of the triangle area formula, area = 0.5 × base × height, that the largest area attainable by a triangle with two fixed sides is one-half the product of the two side lengths, which occurs when the two sides are perpendicular. So the maximum area of the quadrilateral is
0.5 × PA × PB + 0.5 × PB × PC +
    0.5 × PC × PD + 0.5 × PD × PA = 31.5.

This result occurs when the diagonals of $ABCD$ are perpendicular and P is their intersection point.

31. 8/3.

We can draw radii OG and OF to create similar right triangles AGO and OFB.

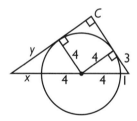

The hypotenuse of △OFB is 4 + 1, or 5, and leg OF is 4. Thus, △OFB is a 3-4-5 triangle, and $FB = 3$. We can find $x = AD$ by comparing side lengths in our similar triangles:

$$\frac{4}{x + 4} = \frac{3}{4 + 1}$$
$$3x + 12 = 20$$
$$x = \frac{8}{3}.$$

32. 3.15 inches; 57.7 percent.

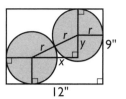

From the diagram, we can see that $x = 12 - 2r$ and $y = 9 - 2r$. Since $x^2 + y^2 = (2r)^2$, we get $(12 - 2r)^2 + (9 - 2r)^2 = (2r)^2$. Therefore, $4r^2 - 84r + 225 = 0$. Using the quadratic formula, we get

$$r = \frac{84 \pm \sqrt{3456}}{8}.$$

Since $r < 12$, the only feasible root is

$$r = \frac{84 - \sqrt{3456}}{8}.$$

To the nearest hundredth of an inch, $r = 3.15$. The area of the sheet cut is $\pi(3.15)^2 \times 2$, or 62.34 square inches. The total area of the sheet is 9″ × 12″ = 108 square inches.

$$62.34/108 \approx 57.7\%$$

33. 3.

Let $AE = x$. The area is $2(10 \cdot 15 + 10x + 15x)$. The volume equals $10 \cdot 15 \cdot x$.
$$2(150 + 25x) = 150x$$
$$x = 3$$

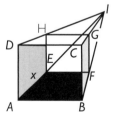

34. 84 square units.

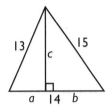

Drawing an altitude to the triangle yields three equations:

(1) $\qquad a + b = 14$

(2) $\qquad a^2 + c^2 = 169$

(3) $\qquad b^2 + c^2 = 225.$

Subtracting equation (2) from equation (3) gives $b^2 - a^2 = 56$. Factoring $b^2 - a^2$ gives $(a + b)(b - a) = 56$. From equation (1), $a + b = 14$, so we have $14(b - a) = 56$, or $b - a = 4$. Adding this equation to equation (1) gives $b - a + a + b = 14 + 4$, or $2b = 18$, or $b = 9$. From equation (3), we get $81 + c^2 = 225$, or $c^2 = 144$, or $c = 12$ (since $c > 0$). The area of the triangle is $(12 \times 14)/2 = 84$.

35. The key is to circumscribe a circle about triangle *ABC*. Extend *BD* to *E*, where it intersects the circumcircle, and construct *CE*. The proof follows: $m\angle ABE = m\angle EBC$ (since *BD* bisects $\angle ABC$) and $m\angle BAC = m\angle CEB$ (both measured by 1/2 arc *BC*).

$$\triangle ABD \sim \triangle EBC;$$

$$\frac{c}{t} = \frac{t + r}{a};$$

$$ac = t^2 + tr;$$

and $xy = tr$ (as products of chords intersecting within a circle); so

$$ac = t^2 + xy.$$

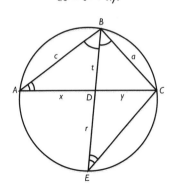

36. $x = 2\sqrt{21}$.

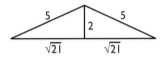

$$h = \sqrt{25 - 4} = \sqrt{21}$$

Area of the first triangle

$$= \frac{1}{2}bh = \frac{1}{2}(4)\left(\sqrt{21}\right) = 2\sqrt{21}.$$

Area of the second triangle

$$= \frac{1}{2}bh = \frac{1}{2}\left(2\sqrt{21}\right)(2) = 2\sqrt{21}.$$

Thus, $\qquad x = 2\sqrt{21}$.

(Adapted from Lehigh University High School Mathematics Contest.)

37. 21.5 degrees.

The situation as described can be *roughly* represented in the following diagram:

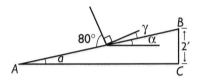

From trigonometry we know that

$$\sin\alpha = \frac{2}{10},$$

so that

$$\alpha = \sin^{-1}\left(\frac{2}{10}\right)$$

$$= 11.5°\left(\text{to the nearest tenth of a degree}\right).$$

We also know that

$$\gamma = 180° - (80° + 90°) = 10°.$$

Thus, the angle that the foot of the cart makes relative to the ground is given by

$$\gamma + \alpha = 10° + 11.5° = 21.5°.$$

38. $57{,}158.3\pi$ cubic units.

This result is obtained by rotating about the side having length 19. Detailed calculations follow:

$$V = \frac{1}{3}\pi r^2 h$$
$$= \frac{\pi \cdot 95^2 \cdot 19}{3}$$
$$= \frac{171{,}475\pi}{3}$$
$$\approx 57{,}158.3\pi$$

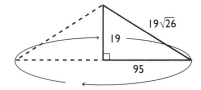

$$V = \frac{\pi \cdot 19^2 \cdot 95}{3}$$
$$= \frac{34{,}295\pi}{3}$$
$$\approx 11{,}431.7\pi$$

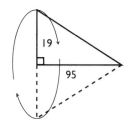

$$V = \frac{\pi y^2 x}{3} + \frac{\pi y^2\left(19\sqrt{26} - x\right)}{3}$$
$$= \frac{\pi y^2 19\sqrt{26}}{3}$$

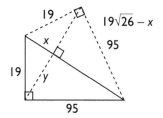

Using the Pythagorean theorem to solve for x and y gives $y = 95/\sqrt{26}$ and $x = 19\sqrt{26}$. Substituting for y gives

$$\pi \cdot \frac{95^2}{26} \cdot \frac{19\sqrt{26}}{3} = \frac{171{,}475\pi}{3} \cdot \frac{\sqrt{26}}{26}$$
$$\approx 11{,}209.7\pi.$$

39. 2 square units.

Observe that $\triangle CGH$ and $\triangle KDH$ are congruent. Therefore, the area of the piece marked x equals the area of a trapezoid $BCGE$, or

$$\left(\frac{1.5 + 2.5}{2}\right)(1).$$

SOLUTIONS TO LOGS AND EXPONENTS

1. $2^{100}, 5^{50}, 3^{75}$.

$$(2^4)^{25} < (5^2)^{25} < (3^3)^{25}$$

2. 41.

$$4^{20} + 4^{20} = 2(4^{20}) = 2(2^{40}) = 2^{41}.$$

3. (c) 9^{x+1}.

$$n^2 = (3^x + 3^x + 3^x)^2$$
$$= (3 \cdot 3^x)^2$$
$$= (3^{x+1})^2$$
$$= (3^2)^{x+1}$$
$$= 9^{x+1}$$

4. $137/20$, or 6.85.

$$\log_2 8 + \log_4 8 + \log_8 8 + \log_{16} 8 + \log_{32} 8$$

$$= 3 + \frac{3}{2} + 1 + \frac{3}{4} + \frac{3}{5} = \frac{137}{20}.$$

5. $\left(\dfrac{1}{\sqrt{2}}\right)^{1000} + \left(1 - \dfrac{1}{\sqrt{2}}\right)^{1000}$.

The function $f(x) = x^{1000} + (1-x)^{1000}$ is a

decreasing function for $0 < x < 1/2$. Since

$$\left(1 - \frac{1}{\sqrt{2}}\right) < \frac{1}{3} < \frac{1}{2}, f\left(1 - \frac{1}{\sqrt{2}}\right) > f\left(\frac{1}{3}\right).$$

Therefore, $\left(1 - \left(1 - \dfrac{1}{\sqrt{2}}\right)\right)^{1000} + \left(1 - \dfrac{1}{\sqrt{2}}\right)^{1000} >$

$\left(\dfrac{1}{3}\right)^{1000} + \left(\dfrac{2}{3}\right)^{1000}$.

Note that considering a simpler case initially might be helpful. For example, suppose the exponents were 2s instead of 1000s.

6. 39.

1 km = 1,000,000 mm.

400,000 km = 4×10^{11} mm.

∴ we need to find the smallest n for which
$2^n > 4 \times 10^{11}$.

∴ $n \log 2 > \log(4 \times 10^{11})$

∴ $n > 38.5$.

7. $x = \{1, 2, 3, 4, 5\}$.

Either $x^2 - 9x + 20 = 0 \quad \Rightarrow \quad x = 4, 5$

or $x^2 - 5x + 5 = 1 \quad \Rightarrow \quad x = 1, 4$

or $x^2 - 5x + 5 = -1$ and $x^2 - 9x + 20$ is even

$\Rightarrow \quad x = 2, 3$.

8. 5.

Substitute 2^x for 15 in the second equation:

$$(2^x)^y = 32,$$
$$2^{xy} = 32$$
$$xy = 5.$$

9. $2/3$.

$$8^{3x+1} - 8^{3x} = 448$$
$$8^{3x}(8 - 1) = 448$$
$$8^{3x}(7) = 448$$
$$8^{3x} = 64$$
$$3x = 2$$
$$x = 2/3$$

10. -16.

$$16^{x^2 + x + 4} = 32^{x^2 + 2x}$$
$$(2^4)^{x^2 + x + 4} = (2^5)^{x^2 + 2x}$$
$$(2)^{4x^2 + 4x + 16} = (2)^{5x^2 + 10x}$$
$$4x^2 + 4x + 16 = 5x^2 + 10x$$
$$x^2 + 6x - 16 = 0$$

The product of the roots is -16.

11. 3.

$$\left(\sqrt[3]{\sqrt{30} + \sqrt{3}}\right) \cdot \left(\sqrt[3]{\sqrt{30} - \sqrt{3}}\right) = \sqrt[3]{\left(\sqrt{30}\right)^2 - \left(\sqrt{3}\right)^2}$$

$$= \sqrt[3]{27} = 3.$$

12. 3/8.

$$\left\{\frac{a}{b}\left[\frac{b}{a}\left(\frac{a}{b}\right)^{1/2}\right]^{1/2}\right\}^{1/2} = \left(\frac{a}{b}\right)^{k}$$

$$\left\{\frac{a}{b}\left[\left(\frac{a}{b}\right)^{-1}\left(\frac{a}{b}\right)^{1/2}\right]^{1/2}\right\}^{1/2} = \left(\frac{a}{b}\right)^{k}$$

$$\left\{\frac{a}{b}\left[\left(\frac{a}{b}\right)^{-1/2}\right]^{1/2}\right\}^{1/2} = \left(\frac{a}{b}\right)^{k}$$

$$\left[\left(\frac{a}{b}\right)^{1}\left(\frac{a}{b}\right)^{-1/4}\right]^{1/2} = \left(\frac{a}{b}\right)^{k}$$

$$\left[\left(\frac{a}{b}\right)^{3/4}\right]^{1/2} = \left(\frac{a}{b}\right)^{k}$$

$$\left(\frac{a}{b}\right)^{3/8} = \left(\frac{a}{b}\right)^{k}$$

Thus, $k = 3/8$.

13. 8.

Since $\log_b(xy) = 11$, then $\log_b x + \log_b y = 11$. Similarly, $\log_b(x/y) = \log_b x - \log_b y = 5$. Adding the two equations gives $2\log_b x = 16$, and so $\log_b x = 8$.

14. 3844.

Alternately square and simplify the equation:

$$1 + \sqrt{3 - \sqrt{1 + \sqrt{2 + \sqrt{x}}}} = 1$$
$$\sqrt{3 - \sqrt{1 + \sqrt{2 + \sqrt{x}}}} = 0$$
$$3 - \sqrt{1 + \sqrt{2 + \sqrt{x}}} = 0$$
$$\sqrt{1 + \sqrt{2 + \sqrt{x}}} = 3$$
$$1 + \sqrt{2 + \sqrt{x}} = 9$$
$$\sqrt{2 + \sqrt{x}} = 8$$
$$2 + \sqrt{x} = 64$$
$$\sqrt{x} = 62$$
$$x = 3844.$$

15. $a = 5$, and $b = 2$.

Observe that $9^1 = 9$, $9^2 = 81$, $9^3 = 729$, and $9^4 > 3000$. Hence, $b < 4$, $a \neq 0$, and $(2^a)(9^b)$ is even. Therefore, $b = 0$ or 2. If $b = 2$, we get $(2^a)(81) = 2a92$. We can check that $a = 5$ satisfies this equation. The case of $b = 0$ is impossible because it would require a power of 2 to end in 0.

16. $(x, y) = \{(16, 243), (81, 32)\}$.

We require that $A + B = 5$ and $A^3 + B^3 = 35$ where $A = x^{1/4}$ and $B = y^{1/5}$. By inspection, we get $A = 2$ and $B = 3$ or $A = 3$ and $B = 2$. Hence, $x = 16$ and $y = 243$ or $x = 81$ and $y = 32$.

17. 4.

$$P^2 - Q^2 = (P - Q)(P + Q)$$
$$= (2 \cdot 2^{-1988})(2 \cdot 2^{1988}) = 4.$$

18. $3M + 2P - N$.

$$78.4 = \frac{784}{10}$$
$$= \frac{392}{5}$$
$$= \frac{8 \cdot 49}{5}$$
$$= \frac{2^3 \cdot 7^2}{5}$$

Therefore,

$$\log(78.4) = \log\left(\frac{2^3 \cdot 7^2}{5}\right)$$
$$= \log 2^3 + \log 7^2 - \log 5$$
$$= 3\log 2 + 2\log 7 - \log 5$$
$$= 3M + 2P - N.$$

19. 6.

Note that $\log_a(c \cdot d) = \log_a c + \log_a d$. We can use this equation to reconstruct the table, taking, for example, $\log_a 6 = \log_a 2 + \log_a 3$; in other words, we can factor each number and then write its log as the sum of the logs of its factors. This example is possible, as all primes less than 10 are given, and we know that $\log_a 1 = 0$ for any number a. Then we get the following table:

n	$\log_a n$
1	0
2	0.387
3	0.613
4	0.774
5	0.898
6	1.000
7	1.086
8	1.161
9	1.226
10	1.285

Since $\log_a 6$ is about 1 and $\log_a a = 1$, we conclude that a is approximately 6.

20. 1/5.

We find that $a = 3, b = 3/2, c = 3, d = 25,$ $e = 2,$ and $f = 3$. The probability of a 3 occurring on both choices

$$= \frac{1}{2} \times \frac{2}{5} = \frac{1}{5}.$$

21. 729.

If the nth term is a_n, then $a_6 = a_1 + 5d$, where d is the common difference. Hence,

$d = \log_2 9 - \log_2 3 = 2\log_2 3 - \log_2 3 = \log_2 3,$
so $x = a_6 = 6\log_2 3 = \log_2 3^6$.

Therefore, $2^x = 3^6 = 729$.

22. 2185.

Since
$$5^{5^5} = 5^{3125}$$
and
$$\log 5^{5^5} = 3125\,(\log 5) \approx 3125\,(0.69897)$$
$$\approx 2184.3,$$

2185 digits can be found.

(Adapted from *Saint Mary's College Mathematics Contest Problems* [1972].)

23. Placing N root signs in the form

$$\sqrt{\sqrt{\ldots\sqrt{\sqrt{2}}}}$$

produces an expression equal to $(2^{1/2})^N$, which may be rewritten as $2^{2^{-N}}$.

$$-\log_2 \log_2 \sqrt{\sqrt{\ldots\sqrt{\sqrt{2}}}} = -\log_2\left[\log_2\left(2^{2^{-N}}\right)\right]$$
$$= -\log_2\left(2^{-N}\right)$$
$$= -(-N)$$
$$= N$$

SOLUTIONS TO SEQUENCES AND PATTERNS

1. 120.

    Notice that the numbers on the far right of each row are perfect squares. Thus, 100 is at the end of one row, and 121 will be at the end of the next row but one place further to the right. The number below 100 is less than 121, or 120.

2. 63rd and 64th positions.

    The digits 6 and 3 will appear in that order as the ones' digit of 36 and the tens' digit of 37. The 6 will be in the $(9 \times 1) + (27 \times 2)$ or 63rd position.

3. Row 11, column 973.

    The initial entries in the respective rows are 1, 2, 4, 8, ..., 2^{n-1}, where n is the row number. $1024 < 1996 < 2048$, or $2^{10} < 1996 < 2^{11}$, so 1996 is in row 11. The column is $1996 - 1024 + 1$, or 973.

4. $n^2 + 3n + 2$.

5. 761.

    One visual representation for the pattern is $n^2 + (n - 1)^2$ as shown in this regrouping.

    Other equivalent representations for the nth term can be found.

6. 129.

    One solution is obtained by adding the number directly above, that directly left, and that diagonally above and left. Therefore the missing number is $52 + 52 + 25 = 129$.

7. 1024/243.

    The sides of a 30°-60°-90° triangle have a ratio

    $$\frac{1}{2}a : \frac{\sqrt{3}}{2}a : a$$

    where a is the length of the hypotenuse.

Hence, for the first triangle,

$$\left(\frac{\sqrt{3}}{2}\right)a = 1, \text{ so } a = \frac{2}{\sqrt{3}};$$

for the second triangle,

$$\left(\frac{\sqrt{3}}{2}\right)a = \frac{2}{\sqrt{3}}, \text{ so } a = \left(\frac{2}{\sqrt{3}}\right)^2;$$

for the tenth triangle,

$$a = \left(\frac{2}{\sqrt{3}}\right)^{10} = \frac{2^{10}}{3^5}.$$

8. 4.

    We require that

    $$(n/2)(20 + 160) = 180(n - 2)$$
    $$\therefore n = 4.$$

9. 5.

    If the nth term is 21, the first term is 3, and the common difference is d, then

    $$21 = 3 + (n - 1)d.$$

    Thus, $d(n - 1) = 18$, and we can obtain possible values of n and d by factoring 18. We find $d = 1, 2, 3, 6, 9, 18$, with corresponding values for n of 19, 10, 7, 4, 3, 2, respectively. Only the first five values of n satisfy the requirements of the problem.

10. 5.

    On close examination of this sequence, 1, 3, 2, −1, −3, −2, 1, 3, 2, −1, −3, −2, ... , one can see that the sequence repeats every six terms and that the sum of the first six terms is 0. The sum of the first 16×6, or 96, terms is also 0. Hence, the sum of the first one hundred terms would be $1 + 3 + 2 + (-1) = 5$.

    (Adapted from the senior high school's Calc-U-Solve local contest at the Allegheny Intermediate Unit, Pittsburgh, Pennsylvania, 1990.)

11. No.

    The child will always overshoot the starting point because after n jumps, the distance from the starting point to the child's location is at most $1 + 2 + 4 + ... + 2^{n-1} = 2^n - 1$ feet. Since the $(n + 1)$th jump is 2^n feet, it can never return the child exactly to the starting point.

12. 7.

Note that 189 digits are required to write 1, 2, 3, …, 99 and that $(1996 - 189) \div 3 = 602.3$. So the 1996th digit is the first digit of 702.

13. 4, 317, 12, 4, 317, 12, 4, 317, 12, 4, 317, 12.

By letting the fifth integer be A, we can see that the following sequence emerges for the first twelve integers:

(___, ___, ___, 4, A, 329 − A, 4, A, 329 − A,
4, A, 329 − A).

But $329 - A = 12$. Hence, $A = 317$. Working backward, we can fill in the remaining figures.

14. $A = 6, B = 7, C = 2$.

By inspecting the ones' digits, we see that $C = 2$. Then $A = 6$ because $32 - 03 = A1 - 32$.
∴ $B = 7$, since $B03 - 6B4 = 29$.

15. $(-1, 2)$.

Try some different examples like $2x + 3y = 4$ and $2x - 7y = -12$. Then prove the result in general.

16. $2i + 95$.

$$i^{0!} + i^{1!} + i^{2!} + i^{3!} + i^{4!} + i^{5!} + \ldots + i^{100!}$$
$$= i^1 + i^1 + i^2 + i^6 + i^{24} + i^{120} + \ldots + i^{100!}$$
$$= i + i - 1 - 1 + \underbrace{1 + 1 + \ldots + 1}_{\substack{\text{97 terms with} \\ \text{exponents that are} \\ \text{multiples of 4}}}$$
$$= 2i + 95.$$

17. $8 - 2\pi$.

The area of the first, or outermost, shaded region is $4 - \pi$. A side of the second square is $\sqrt{2}$, and the area of its shaded region is $(4 - \pi)(1/2)$. This pattern continues, giving the infinite series

$$\left(4 - \pi\right) + \frac{1}{2}\left(4 - \pi\right) + \frac{1}{4}\left(4 - \pi\right) + \ldots$$
$$= \left(4 - \pi\right)\left(1 + \frac{1}{2} + \frac{1}{4} + \ldots\right)$$
$$= \left(4 - \pi\right) \cdot 2 = 8 - 2\pi.$$

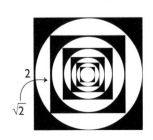

18. $45 + 1/\sqrt{2}$.

Remember that $\sin A° = \cos (90° - A°)$ and that $\sin^2 A + \cos^2 A = 1$. Thus,

$$\sin^2 1° + \sin^2 2° + \sin^2 3°$$
$$+ \ldots + \sin^2 44° + \sin^2 45° + \sin^2 46°$$
$$+ \ldots + \sin^2 89° + \sin^2 90°$$

is equivalent to

$$\sin^2 1° + \sin^2 2° + \sin^2 3°$$
$$+ \ldots + \sin^2 44° + \underline{\sin^2 45°} + \cos^2 44°$$
$$+ \ldots + \cos^2 1° + \underline{\sin^2 90°}.$$

This simplifies to

$$44 + 1/\sqrt{2} + 1 = 45 + 1/\sqrt{2}.$$

19. 7.

Let $t_n = a + (n - 1)d$, where t_n is the nth term, a represents the first term, and d is the constant difference. Hence,

$$t_{10} = a + 9d = 64$$
$$d = \frac{64 - a}{9}.$$

Since the terms are whole numbers, $(64 - a)$ must be a multiple of 9. Since $d > 0$, $a = 1, 10, 19, 28, 37, 46,$ or 55, and the corresponding values for d can be calculated.

20. $(s - 1)^2$ square units.

Consider table 1. Observe that the values for the unshaded areas are perfect squares. In fact, in general they represent $(s - 1)^2$.

Suppose that we extended the table. For any value of s, $n = 2s - 1$. Hence, the unshaded area equals $s^2 - (2s - 1)$, or $s^2 - 2s + 1 = (s - 1)^2$ square units.

Table 1

	Shaded and Unshaded Areas	
Length of Side of Large Square (s)	Number of Identical Shaded Squares (n)	Unshaded Area
1	1	0
3	5	4
5	9	16
7	13	36

21. Since one entry is in the first row, two are in the second row, three are in the third, and so on, we can use the summation formula

$$1 + 2 + 3 + \ldots + n = \frac{n(n+1)}{2}$$

to find the number of rows involved. With some trial and error, we see that when n equals 20, 210 terms occur. Hence, we are summing the entries in the first twenty rows, along with the first two entries in the twenty-first row. The sum of the entries in the first twenty rows of Pascal's triangle is given by $2^0 + 2^1 + 2^2 + \ldots + 2^{19}$, a geometric series that sums to $2^{20} - 1$. Since the first two entries in the twenty-first row are 1 and 20, the sum of all 212 terms is given by $(2^{20} - 1) + (1 + 20) = 2^{20} + 20$; hence, $k = 20$.

Note that $n(n + 1)/2 = 210$ implies that $n(n +1) = 420$. Therefore, the value of n can be determined using the quadratic formula. It is more elegant to observe that the value of $n = [\sqrt{420}]$, where $[x]$ represents the greatest integer less than or equal to x. This fact holds true because n is the smaller of a pair of consecutive positive integers that have a product of 420.

22. 72.
If each team played each of the other teams once, a total of 10 choose 2, or $10!/8!2! = 45$, games would be played. Since the teams play each of the other teams eighteen times, a total of $45 \cdot 18 = 810$ games are played. Since each game results in a win, a total of 810 wins occur during the season. Let w be the number of games that the last-place team won and c be the constant difference in wins between consecutive teams. Then

$$w + (w + c) + (w + 2c) + \ldots + (w + 9c) = 810,$$
$$10w + 45c = 810,$$
$$2w + 9c = 162.$$

To make w as large as possible, c must be as small as possible. If $c = 1$, w is not an integer. If $c = 2$, $w = 72$. Therefore, the largest number of games that the last-place team could have won is 72.

23. 1/64 cup.
Each time a half-cup of liquid is consumed, half the hot chocolate in the mixture disappears. Thus, the amounts of hot chocolate consumed in each of the six half-cups are 1/2, 1/4, 1/8, 1/16, 1/32, and 1/64 cup, respectively. The remaining hot chocolate is 1/64 cup, since half the hot chocolate is consumed each time.

24. $(n + 1)! - 1$.
We can rewrite $1(1!) + 2(2!) + 3(3!) + \ldots + n(n!)$ as

$2(1!) + 3(2!) + 4(3!) + \ldots + n[(n - 1)!] + (n + 1)(n!)$
$- 1! - 2! - 3! - \ldots - (n - 1)! - n!$

This expression simplifies to

$$(n + 1)(n!) - 1 = (n + 1)! - 1.$$

Note that $2(1!) = 2!, 3(2!) = 3!$, and so on. Hence, the terms in the sequence represented as $2(1!) + 3(2!) + \ldots + (n + 1)n!$ will cancel out with the terms one over to the right below them except for $(n + 1)n! - 1!$.

25. $k = 925$.
Let x be the second term of the sequence and a be the difference between successive terms; then the first three terms are $x - a, x, x + a$, and the following equations result:

$$\text{I: } (x - a)^2 = 36 + k,$$
$$\text{II: } x^2 = 300 + k,$$
$$\text{III: } (x + a)^2 = 596 + k.$$
$$\text{III} - \text{II: } a(2x + a) = 296$$
$$\text{II} - \text{I: } a(2x - a) = 264$$

Subtracting the last equation from the preceding one, we get

$$2a^2 = 32;$$
$$\therefore a = \pm 4.$$

Substituting, we get

$$x = \pm 31.$$

26. 27,000,001.
Pair as follows:

0	999,999
1	999,998
2	999,997
⋮	⋮
499,999	500,000

These 500,000 pairs each total 999,999. The digit sum of each pair is 54. Thus 500,000 × 54 = 27,000,000. Then 1,000,000 contributes a digit sum of 1 to the total, making 27,000,001.

SOLUTIONS TO COUNTING

1. 29.

Without using pennies, we can make 5, 10, 15, ..., 45 cents. By using the pennies as well, we can make 1, 2, 6, 7, 11, 12, ..., 46, 47 cents.

2. 48.

There are $6 \times 4 \times 2$ ways to place the numbers 1, 2, and 3 on faces. The positions of 4, 5, and 6 are determined by their opposite counterpart.

3. 13.

The following Venn diagram represents the given information. In the diagram, T = tibbs, G = gibbs, and P = pibbs.

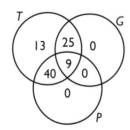

4. 5.

In addition to AED, the following angles are acute: AEB, AEC, BEC, BED, and CED.

5. 2.

The total number of handshakes, 68, equals $n(n-1)/2 + k$, where $k < n$ and k represents the number of people present known by the straggler. The only plausible value of $n(n-1)/2$ is 66. Hence, $k = 2$.

6. 8.

The following diagram shows the various solutions. Note that each of the three original circles can be internally or externally tangent to the fourth circle. This fact accounts for the $2 \times 2 \times 2 = 8$ arrangements of the circles.

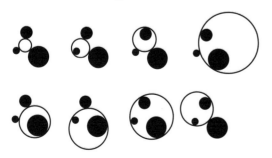

7. 279.

There are 9×10 numbers of the form AB75, 10×10 numbers of the form 75AB, and 9×10 of the form A75B. Since 7575 is counted twice, the result is $90 + 100 + 90 - 1$.

8. 32.

The Venn diagram shows that 16 people do not recycle at least one of the items. Therefore, $48 - 16$ recycle all three.

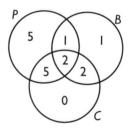

9. 360.

Sixty times per hour for 10, 11, 12, and 1. Fifteen times per hour for the remaining eight hours.

10. 69.

The smallest number in the set is 1023. Exactly fifty-six numbers begin with 10, since we may choose the third digit in eight ways and the fourth digit in seven ways. Next, we get the seven numbers from 1203 through 1209, inclusive. Then we reach the numbers 1230, 1234, 1235, 1236, 1237, and 1238.

11. 9120.

The members can be selected in $40 \times 38 \times 36$ ways. Since order does not matter, we must divide by 3!.

12. 217.

Consider the integers from 200 through 299. All but nineteen do not have a 3. Consider the integers from 300 through 399. None of these integers satisfies the requirements.

For each interval 0 through 99, 100 through 199, 400 through 499, 500 through 599, 600 through 699, 700 through 799, 800 through 899, and 900 through 999, seventeen integers satisfy the given requirements. Therefore, the total number of cases is $81 + 0 + (8 \times 17)$, or 217.

13. 42.

Consider the following diagram. The intersections, or vertices, on the map are represented by circles. The numbers in the circles equal the number of different paths that one can take from A to that intersection. Observe how the number of paths can readily be calculated by summing the numbers for the intersections at which one could have been at the preceding step.

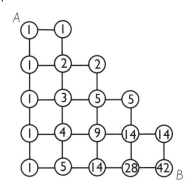

14. The eight possibilities and their areas are shown:

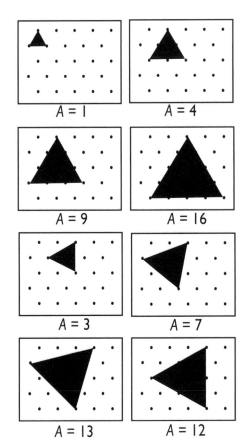

15. 156.

Consider the offices to be numbered 1, 1, 2, 2, 3, 4. Without a loss of generality, we can let the six people be lined up in some fixed order with the first two being those who refuse to share an office. The number of distinguishable permutations of 112234 is the total number of ways without any restrictions. Subtract from this total the number of permutations that begin with either 11 or 22, since those indicate that the first two people are assigned to double offices 1 or 2. The number of ways is

$$\frac{6!}{2!\,2!} - \frac{4!}{2!} - \frac{4!}{2!} = 180 - 12 - 12.$$

16. 24.

There are $\binom{7}{3}$ = 35 possible sets of vertices.

However, $\binom{5}{3}$ = 10 sets could be chosen from the line containing A, B, C, D, and E, and another $\binom{3}{3}$ = 1 could be chosen from that containing C, F, and G. Hence, 35 − 10 − 1 possible triangles can be made.

17. 75.

We can find

$$\binom{9}{3} = 84$$

possible sets of three points, nine of which are collinear. Therefore, 84 − 9 = 75 triangles are possible.

18. 75.

Dimensions	Number of Rectangles
1 x 1	15
1 x 2	21
1 x 3	13
1 x 4	5
2 x 2	7
2 x 3	8
2 x 4	3
3 x 3	2
3 x 4	1
	75

19. 156.

The only one-digit multiple of 4 would be 4. The two-digit multiples of 4 would be restricted to 12, 24, 32, 44, and 52. These same numbers—namely, 12, 24, 32, 44, and 52—would be the "endings" of larger numbers that are multiples of 4. Hence, $5 \times 5 = 25$ three-digit numbers and $5 \times 5 \times 5 = 125$ four-digit numbers would be allowed. Note that the hundreds digits, the thousands digits, or both, can be any of 1, 2, 3, 4, 5. The total number of acceptable multiples of 4 is given by $125 + 25 + 5 + 1$.

20. 13.

(Adapted from Bruni and Silverman [1977] and Lindquist [1977].)

21. 300.

The first digit must be a 5 or 6. The number of distinct arrangements greater than 5,000,000 equals

$$\frac{5 \cdot 6!}{3! \cdot 2!}.$$

22. 27.

No such single-digit numbers exist. Suppose that such a number has two digits. It must be even, since 8 is even, so the units digit is either 2 or 4. If the units digit is 2, a quick check shows that only 32 works. If the units digit is 4, only 24 works. So two such numerals contain two digits.

Suppose that such a number has three digits. A multiple of 8 must be a multiple of 4. We can check for multiples of 4 by looking at the rightmost two digits: if they form a multiple of 4, the entire number is a multiple of 4. This requirement means that the rightmost two digits are 12, 24, 32, or 52. Taking each case separately, we find only 312, 512, 432, 152, and 352. Five such numerals contain three digits.

Continue to count the four- or five-digit numerals. Since 1000 is a multiple of 8, any four- or five-digit numeral ending in 312 is of the form $1000n + 312$ and hence a multiple of 8. Therefore, 4312, 5312, 45312, and 54312 are all satisfactory. Similarly, each of 512, 432, 152, and 352 will produce four more such numbers, thus giving $5 \times 4 = 20$ four- or five-digit numbers. The total of satisfactory numbers is $2 + 5 + 20$, or 27.

23. 867.

Listing and counting would not be practical for solving this problem. Some kind of clever strategy is needed.

Start with an N, since the array has only four, and note the seventeen readings of NAH—backward, forward, diagonally, and so on. Hence, seventeen readings of HAN are possible for each N.

Take one specific N and consider its seventeen associated HANs. Each HAN can be linked to three other Ns; each N has seventeen NAHs. Hence, the number of ways to spell HANNAH is $17 \times 3 \times 17 = 867$.

SOLUTIONS TO WORD PROBLEMS

1. 4000.
   Each tire was used for 4/5 of the distance, or 4000 miles.

2. 18 correct, 5 wrong, 2 missing.
   A perfect test score would be 125. For each wrong answer, w, subtract $(5 + 4 = 9)$; for each missing answer, m, subtract $(5 + 3 = 8)$. Thus,

   $$125 - 9w - 8m = 64$$
   $$9w + 8m = 61$$
   $$w = \frac{61 - 8m}{9}.$$

   For w to be integral, $m = 2$. Thus, $w = 5$ and $c = 18$.

3. 72.
   Solve for n where

   $$\sqrt{\frac{n}{2}} + 2 = \frac{n}{9}$$

   and n is an integer.

   $$\therefore \frac{n}{2} = \left(\frac{n-18}{9}\right)^2$$

   Multiplying both sides by 162 and simplifying give

   $$2n^2 - 153n + 648 = 0$$
   $$\therefore (2n - 9)(n - 72) = 0.$$

   (Adapted from Devi [1990].)

4. 37/125 mile.
   Tom runs 50/10 miles in the time it takes Dick to run 49/10 miles. Also, Dick runs 25/5 miles in the time it takes Harry to run 24/5 miles. Hence, Harry would run (24/25)(49/50)(5) miles or 588/125 miles in the time it takes Tom to run 5 (or 625/125) miles.

5. The shorter jar.
   If the volume of the taller jar is $\pi r^2 h$, then the volume of the shorter jar is $\pi(2r)^2 (h/2)$, or $2\pi r^2 h$. It is worth paying 1.5 times the price for twice the volume.

6. 40.
   The ratio of the penalties equals the ratio of the excess weights.

   $$\therefore \frac{6.50}{2.50} = \frac{105 - x}{105 - 2x}$$
   $$\therefore \quad x = 40$$

   (Adapted from Billstein, Libeskind, and Lott [1993].)

7. Vic is 53 years old. He has 2 children and a 303-foot yacht, or 3 children and a 202-foot yacht, or 6 children and a 101-foot yacht.
   $$32,118 = 2 \times 3 \times 53 \times 101$$

   (Adapted from Pólya and Kilpatrick [1974].)

8. 76.
   A total of 20 men × 8 days = 160 man-days to do one-fourth of the job. Three-fourths of the job remain. The man-days that remain are $3 \times 160 = 480$, and 480 man-days divided by 5 days = 96 men. Hence, 76 additional men were required.

   (Problem and solution are from Whimbey and Lockhead [1984].)

9. Half the time.
   If each man walks D miles and then rides D miles in time T,

   $$\frac{D}{4} + \frac{D}{12} = T,$$

   which means that $D = 3T$. Since they cover $2D$ or $6T$ miles in T hours, they progress at a rate of 6 MPH. The horse's speed is 12 MPH; therefore, the horse rests half the time.

   (From Dunn [1980].)

10. **22.6 minutes.**

The total volume of water is $\pi(4^2)(12.5) = 200\pi$. Let x represent the depth of water when the depth is equal in both tanks. At that time, the volume of water in the tank with radius 4 is $16\pi x$, and the volume of water in the tank with radius 3 is $9\pi x$:

$$16\pi x + 9\pi x = 200\pi$$
$$25\pi x = 200\pi$$
$$x = 8.$$

The amount of water that has been transferred is $\pi(3^2)(8) = 72\pi$. The time required to pump this much water is $(72\pi)/10$, or approximately 22.6 minutes.

11. **29 chih.**

Take $7 \times 3 = 21$ as one side of a right-angle triangle. The height of the tree, 20, is another side. The length of the vine equals $\sqrt{21^2 + 20^2}$, or 29, chih.

12. **$15,400,000.**

The highest possible salary (in millions of dollars) would equal:

$$20 - (1 + 13 \times .2 + 10 \times .1), \text{ or } 15.4.$$

In fact, this makes the pitcher the highest paid player.

(From Grossman [1990].)

13. **Joe has 48 walnuts, Mia has 15, and Pepe has 13.**
Let J, P, and M represent the number of walnuts that Joe, Pepe, and Mia have, respectively. Then $P + 10 + 15 = J - 10 + M - 15$, so $P = J + M - 50$. Also, $M + 15 + 8 = J - 15 + P - 8$, so $M = J + P - 46$. Substituting for P, we find that $J = 48$. Substituting $J = 48$ in either equation leads to $P = M - 2$. Observe that Mia has at least 15 walnuts, since she can give away that many. Therefore, Pepe has at least $15 - 2$, or 13, walnuts. The three values $J = 48$, $P = 13$, and $M = 15$ satisfy the original conditions.

14. **$49.99.**

Let c and d be the number of cents and dollars that Elise had when she started out. Then $(100d + c)/2 = 100(c/2) + d$, which simplifies to $98d = 99c$. We know that c and d are whole numbers and that c is less than 100. So the only solution is $d = 99$ and $c = 98$. Hence, Elise started with $99.98 and spent half of it, or $49.99.

(Adapted from Gardner [1958].)

15. **660 feet.**

For a train to pass completely through a tunnel, it must travel the length of the tunnel plus one train length. The time is measured starting when the engine enters the tunnel and ending when the last car leaves the tunnel. The train is 15 seconds long, and the train plus the tunnel is 45 seconds long. Thus the tunnel is two train lengths long, and the train is $(1/2)(1320)$, or 660, feet long.

16. **30 MPH.**

Since the train travels 660 feet in 15 seconds, it would travel 4(660), or 2640, feet in one minute. Since 2640 feet is one-half mile, the train travels one-half mile per minute, or 30 MPH.

17. **2π meters.**

The straight parallel sides are the same length for all runners. The distance around the circular portions of the track for the outside lane is $2\pi r$. The distance around the circular portions of the track for the second lane from the outside is $2\pi(r - 1)$. The difference is 2π, or $2\pi r - 2\pi(r - 1)$. Therefore, the runner on the outside lane should have a lead of 2π meters.

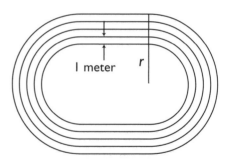

18. **15.**

Let's attempt to find out how many weeks it would take one bulldozer to do the job alone.

Let w = the number of weeks for the bulldozer to complete the job. Let n = the number of bulldozers that the engineer originally had. One bulldozer can complete $1/w$ of the job in one week. According to the statement of the problem, $n + 2$ bulldozers can complete the job in one week. Hence,

$$(n + 2)\frac{1}{w} = 1.$$

Also, since n bulldozers can complete the job in 8 days, this number of bulldozers can complete 7/8 of the job in 1 week. Hence,

$$n\left(\frac{1}{w}\right) = \frac{7}{8}.$$

From the first equation, we get

$$\frac{n}{w} + \frac{2}{w} = 1.$$

From the second equation, we get

$$\frac{n}{w} = \frac{7}{8}.$$

Substituting this value for n/w in the first equation, we have

$$\frac{7}{8} + \frac{2}{w} = 1,$$
$$\frac{2}{w} = \frac{1}{8},$$
$$w = 16.$$

Thus, one bulldozer can do the job in 16 weeks. Note that the problem asks how many weeks the job would be behind schedule. The engineer first said the job could be completed in a week.

Alternatively, if n is the original number of bulldozers, then $n + 2$ bulldozers can do 1/7 of the job in one day. Therefore, two bulldozers can do $1/7 - 1/8 = 1/56$ of the job in one day, and hence one bulldozer can do 1/112 of the job in one day. Thus, one bulldozer takes 112 days (16 weeks) to do the job alone.

19. The pencil is $0.26, the eraser is $0.19, and the notebook is $0.55, assuming integer numbers of cents for each item.
A graphical solution is provided here. A detailed algebraic solution also appears in the original calendar solutions.

The givens are $p + e + n = 1$, $n > 2p$, $3p > 4e$, and $3e > n$. So $n = 1 - e - p$; substitute into the other equations for

$$1 - e - p > 2p \Rightarrow 1 - e > 3p, \quad 3p > 4e,$$

and

$$3e > 1 - e - p \Rightarrow 4e > 1 - p.$$

Graphing this version on e versus p, one gets the following graphical solution.

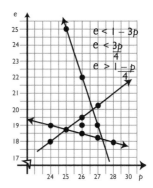

The only integral solution for p and e in the interior of the triangle is $p = 0.26$, $e = 0.19$, and thus $n = 0.55$.

(Adapted from Dunn [1983].)

1. One possible solution appears in the following diagram:

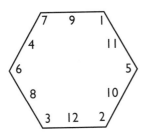

Hint: All numbers follow one number and precede another number except for 1 and 8, which start and end the list. Put 1 and 8 in the boxes with the most neighbors—that is, in the center.

(From Ecker [1987].)

2. One arrangement is given; other possibilities exist.

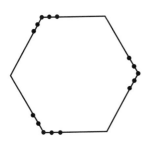

3. Place dots at alternate vertices. This leaves twelve dots to be placed. Hence, we must place two additional dots on each side, not at vertices.

4. 20.

5. $98 - 76 + 54 + 3 + 21 = 100.$
No other solution uses only four plus or minus signs. An extensive collection of solutions to both the ascending and descending sequences appears in Gardner (1967).

6. 8.
A cannot equal—
0, because *M* and *N* would equal 0;
1, because the product *MAN* is different from *AS*;
2, because a three-digit product would not be possible;
3, because 4 cannot be carried to $A \times A$;
4 or 7, because 8 cannot be carried to $A \times A$;
5 or 6, because *S* would have to equal 0, making *N* equal to *S*, or *S* would have to equal 1, making *N* equal to *A*;
9, because then 8 would have to be carried, making *A* equal to *S*.

(Problem and solution from Summers [1968].)

7. The maximum amount is $119.

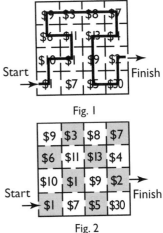

Fig. 1

$9	$3	$8	$7
$6	$11	$13	$4
$10	$1	$9	$2
$1	$7	$5	$30

Start → ... → Finish

Fig. 2

Figure 1 shows that $119 is the maximum by showing that $7 is the minimum amount one can possibly give up. By shading the squares, as in figure 2, one's path must progress from shaded, to unshaded, to shaded, and so on. Starting and ending in shaded squares implies that at least one unshaded square is not part of the path. Omitting the $4 unshaded square would also lead to omitting the $7 shaded square and another unshaded square and, in effect, giving up at least $11. Omitting only the $7 unshaded square was shown to be possible. Thus, the minimum amount one can give up is $7 and the maximum one can collect is $119.

8. $1 \times 26 \times 345 = 8970$ or $2 \times 14 \times 307 = 8596$. (Adapted from Dunn [1983].)

9. The second player.

   If the first player takes one coin, then the second player takes the middle two of the remaining four. That move leaves two isolated coins, but the first player can take only one of them, leaving the final coin for the second player.

   If the first player takes two coins initially, then the second player takes the middle one of the remaining three and wins as before.

10.
$$
\begin{array}{r}
1 \\
333 \\
+ \ 777 \\
\hline
1111
\end{array}
$$

(Problem and solution adapted from Townsend, [1990].)

11.

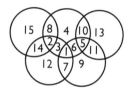

(Problem and solution adapted from Savin [1993].)

12. 2210 in base three is 75 in base ten.

$$
\begin{array}{r}
120 \\
\times \ 12 \\
\hline
1010 \\
120 \\
\hline
2210
\end{array}
$$

13. Yes.

   To win, go first, and take two pennies.

   Keep in mind that if you leave your opponent with the final penny, you will win. It follows that if you leave your opponent five pennies and he or she takes one, two, or three pennies, you can take the necessary amount to leave your opponent with the final penny. Similarly, it can be argued that leaving your opponent with nine or thirteen pennies puts you in a winning position. Thus, you can guarantee a winning position by going first and removing two pennies.

14. Two possible solutions follow. The second solution is shown in parentheses. (*Note:* 8, 9, and 10 must appear in the three squares on the outermost circle.)

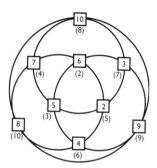

15.

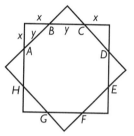

As shown, the regular octagon is *ABCDEFGH*. Note that the points *B* and *C* are not points of trisection of the top horizontal segment—that is, *x* is not equal to *y*.

16. Four cards, no; six cards, yes.

   With four cards, the person who starts is destined to lose. Selecting either one or three cards makes the winning move almost automatic for the second player, who could remove the remaining cards. Suppose that the first player removes two cards. The second player must remove one card because he or she must take a different number of cards from that taken by the previous player. The remaining one card cannot be removed because one card was just removed. Hence, the second player wins.

   With six cards, if the first player removes two cards, the second player is in the position of removing one or three cards. Thus, the first player can remove three cards or one card to win.

17. 10.

The four moves that follow place the first R and W in the correct position.

RRRRWRWWWW
RRRWRRWWWW
RRWRRRWWWW
RWRRRRWWWW

The same reasoning shows that three more moves are needed to place the second R and W, two additional moves are needed for the third R and W, and one move is needed to switch the fourth R and W to achieve the desired result. The total number of moves is $4 + 3 + 2 + 1 = 10$.

18. To prove this situation, we notice that if the two coins are a head and a tail, then by turning them over, the tail becomes a head, the head becomes a tail, and we still have the same number of heads and tails.

If two coins are both heads, then turning them over takes away two heads and adds two tails. Therefore, if the number of heads is even, it stays even. The same holds true if the two coins were originally tails.

Thus, an even number of heads must always be visible, but to get all the coins to land heads, we would need an odd number (3). Therefore, getting all heads is impossible.

(Adapted from Holt [1978].)

19. Seven designs are feasible. The numbers on the faces are indicated here.

First Die	Second Die
{0, 0, 0, 2, 2, 2}	{1, 2, 5, 6, 9, 10}
{0, 0, 0, 3, 3, 3}	{1, 2, 3, 7, 8, 9}
{0, 0, 0, 6, 6, 6}	{1, 2, 3, 4, 5, 6}
{1, 1, 1, 2, 2, 2}	{0, 2, 4, 6, 8, 10}
{1, 1, 1, 3, 3, 3}	{0, 1, 4, 5, 8, 9}
{1, 1, 1, 4, 4, 4}	{0, 1, 2, 6, 7, 8}
{1, 1, 1, 7, 7, 7}	{0, 1, 2, 3, 4, 5}

20. The required pattern of strings can be obtained from the given pattern by reflecting it in the dotted line.

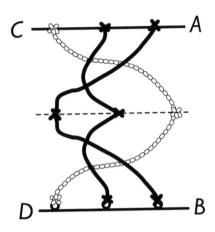

(From Trigg [1985].)

21. (a) 2; (b) 6.

The placement of the queens is shown here for the respective boards. The Os mark the queens, and the Xs mark the unattacked squares.

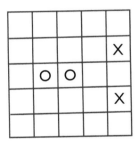

22. Fold the paper in half along the 11-inch side, making a rectangle that is 5 1/2 inches by 8 1/2 inches. Fold one 5 1/2-inch edge from the corner along the 8 1/2-inch side. The uncovered part of the 8 1/2-inch side is exactly 3 inches (8 1/2 − 5 1/2). Alternatively, fold the sheet into a square measuring 8 1/2 inches on a side leaving a 2 1/2-inch strip at the end. Open the sheet and fold over the 2 1/2–inch strip, leaving a length of 6 inches, which can be folded in half.

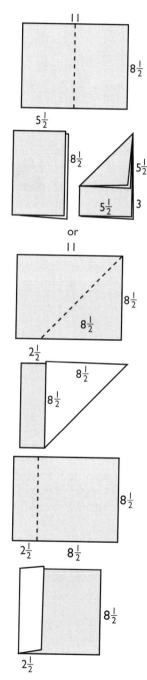

23.

$$
\begin{array}{r}
775 \\
\times\ 33 \\
\hline
2325 \\
2325 \\
\hline
25575
\end{array}
$$

The best approach seems to involve considering a simpler problem—namely, that of identifying values that satisfy the product $PPP \times P = PPPP$. In fact, only four three-digit numbers consist entirely of prime digits that when multiplied by a prime digit produce a four-digit number made up only of prime digits. These products are as follows:

$$775 \times 3 = 2325$$
$$555 \times 5 = 2775$$
$$755 \times 5 = 3775$$
$$325 \times 7 = 2275.$$

Observe that none of the three-digit numbers appears twice in the list. Hence, the multiplier required by the problem must consist of two identical digits. Checking the four possibilities, we find that 775×33 is the only one that equals a five-digit number consisting entirely of prime digits.

(Gardner [1991] credits the creation of this problem to Joseph Ellis Trevor.

24. To solve this problem, such manipulatives as toothpicks would be quite useful. Some of the finished solutions are shown.

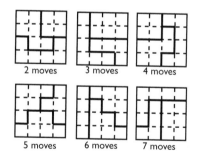

2 moves   3 moves   4 moves

5 moves   6 moves   7 moves

25. Eleven squarelets, as shown.

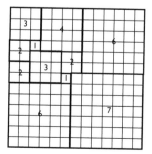

SOLUTIONS TO FACTS

1. Fourth century B.C.

$$A = D = E = 3.$$
$$G = A + 1 = 4.$$

2. 1807.
   (Problem adapted from Osen [1974].)

3. 1740.
   Problem adapted from Osen [1974].)

4. 1805 or 1806.
   The only perfect square in the 1800s is 1849. Thus, De Morgan was 43 years old.

5. The factors of 8128 are as follows: 1, 2, 4, 8, 16, 32, 64, 127, 254, 508, 1016, 2032, and 4064. The sum of the proper factors is 8128.
   Perfect, deficient, and abundant numbers were said to have mystical powers, and their discovery is attributed to the Pythagoreans.

6. The code was devised by Francis Bacon in the seventeenth century to encrypt political secrets. Z would appear as bbaab if Z was the twenty-sixth letter of the alphabet. At the time of Bacon, however, the English alphabet had only twenty-four letters; J and V were not used. Thus, at that time, Z would have appeared as babbb. Thirty-two permutations are possible.

7. The 120th place.
   The first 1 is in the 1! place. The second 1 is in the 2! place. The third 1 is in the 3!, or 6th, place. The fourth 1 is in the 4!, or 24th, place. So the next 1 is in the 5!, or 120th, place.

8. 1749: 1706 + 43.

9. 52 solar years; 73 ritual years.
   Remove the common factor of 5 or find the lowest common multiple to justify the result.

10. 225 sq. units.

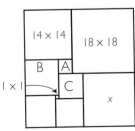

11. 462.

$$V = \pi r^2 h = \pi(7)^2 (3) \approx 462.$$

Consider the diagram where A, B, C, and x are the respective lengths of the sides of the squares marked by those letters.

$$A = 18 - 14 = 4$$
$$B = 14 - A = 14 - 4 = 10$$
$$C = (B + 1) - A = (10 + 1) - 4 = 7$$
$$x = (18 + A) - C = (18 + 4) - 7 = 15$$

Area = x^2 = 225.

12.

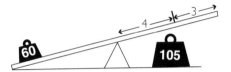

The weight should be placed 4/7 of the distance from the fulcrum to the end of the lever. For the lever to balance, the moments must be equal. A moment is the weight times its distance to the fulcrum. Assume that the 60-pound weight is one unit from the fulcrum. Then

$$60(1) = 105x$$
$$60 = 105x$$
$$\frac{60}{105} = x$$
$$\frac{4}{7} = x.$$

Therefore, the distance from the fulcrum to the 105-pound weight must be 4/7 of the distance from the fulcrum to the 60-pound weight.

13. 4 pounds.
   Since $F = ma$, where F is force, m is mass, and f is the acceleration of gravity, $a = -32$ ft./sec.^{2}, and $F = -2048$ ft. lb./sec.^{2},

$$-2048 = m_A (-32),$$
$$64 = m_A$$

and

$$\left(m_B\right)^3 = m_A = 64$$
$$m_B = 4.$$

14. Letting $n = 10$ gives $a = 3, b = 1$. Hence,

$$\sqrt{10} = 3 + \frac{1}{2(3)+1} = 3\frac{1}{7}.$$

(A reasonable estimate, indeed!)

15.
$$x + \frac{1}{3}x + \frac{1}{(3)(4)}x - \frac{1}{(3)(4)(34)}x =$$
$$= \frac{577}{408}x = 1.414215686x,$$

or the square root of 2 itself. (Is this the value in your calculator for the square root of 2?)

The *Sulvasutras* (c. 500–200 B.C.) were handbooks for the use of Vedic priest-craftsmen in India in building ritual altars according to precise specifications. The title means "Rules of the Cord" and refers to the cords used in measuring. (See Joseph [1991] and van der Waerden [1973].)

16. 55/6.

Observe that the middle share must be 20 (i.e., the mean, or 100/5).

Solving for *d*, where

$(20 - 2d) + (20 - d)$
$\qquad = 1/7\ [20 + (20 + d) + (20 + 2d)],$
we get $d = 55/6$.

17. $a = 1/4; b = 1/4; c = 1/16; d = 1/8; e = 1/16;$
$f = 1/8; g = 1/8.$

Dimensions are shown in the diagram:

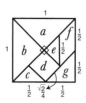

18. 1050, 1200, 2550, 5100.

The Michigan papyrus 620, a Greek papyrus written in the second century, contains three arithmetic problems, including this one. This document's importance lies in its use of a symbol for the unknown quantity, following the method of Diophantus. (See van der Waerden [1983].)

The numbers represented by x, $(8/7)x$, $[x + (8/7)x + 300]$, and $[2x + (16/7)x + 600]$ must total 9900. Hence, $x = 1050$.

SOLUTIONS TO QUICKIES

1. False.

2. 48 MPH.

3. All multiples of 6 (12, 18, 24, 30, ..., 90, 96) and 20, 40, 80, 56, 70, 80, 88.

4. Thirteen.
   Hint: Solve for B: $3B + 1 = 2(B + 7)$.

5. 155.

6. 176.

7. 4999 and 5001.
   Hint: Look for a difference of squares.

8. 75.
   Note that $1.20A = .90C$.

9. 17.
   Note that $493 = 17 \times 29$.

10. 12.
    The integers are 2, 4, and 6.

11. 0, 1, 64, 729, 4096, and 15625.
    Numbers that are both perfect squares and perfect cubes must be sixth powers.

12. 8.
    Note that $23^9 - 23^8 = 23^8(23 - 1)$.

13. 361, 784, and 529.

14. 138 and 777.

15. It is impossible.

16. 1008.

17. 76 m.

18. 5 cents and 8 cents.

19. $6^{200}, 2^{800}, 5^{400}, 3^{600}$.

20. 6.

21. (a) $\dfrac{7293}{14586}, \dfrac{6729}{13458}, \dfrac{7269}{14538}$
    are possible answers.

    (b) $\dfrac{5823}{17469}, \dfrac{5832}{17496}$ are possible answers.

22. $(a, b) = (1, 12), (2, 11), (4, 7), (5, 4)$.

23. $6\sqrt{2}$ units.

24. 1117, 1151, 1171, 1511, 2111.

25. 7.
    Note that the length of the longest side must not exceed 7.

26. 4 hours, 11 minutes. (15:51 to 20:02)

27. No.

28. 29.

29. $\sqrt{9} \times 9 - \sqrt{9}, \left(9 - .\overline{9}\right) \times \sqrt{9}, \sqrt{9}^{\sqrt{9}^{\sqrt{9}}} - \sqrt{9},$
    and $\left(.\overline{9} + \sqrt{\sqrt{9 \times 9}}\right)!$
    are possible expressions.

30. $234 = 3 \times 78$.

31. Yes.
    The altitude from C to AB remains the same but the base is trisected.

32. 27, 64, and 125.

33. 19683.

34. Any triangle obtained by adjoining two right triangles with integer sides will give the desired property.

35. Kramer gets 3/5; Erik gets 2/5.

36. 8 square units.

37. 2 quarters and 2 nickels.

38. 24 cm³.
    Note: $V = LWH = \sqrt{(LW)(WH)(LH)}$ in general.

39. Yes.
    Number the days 1, 2, 3, ..., 365 (366) and list the days corresponding to the thirteenth of each month. These numbers will produce all seven possible remainders on division by 7, thus ensuring a Friday the 13th each year.

40. Four possible solutions are shown here.

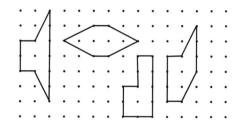

41. 10,000.

42. *b.*
    Note: $a = 2002(2002^2 - 1^2)(2002^2 - 2^2) < b = 2002^5$.

43. 80.

44. 58.

45. 1001, 2002, and 3003.

46. 20.

47. 6.

48. 2 and 997.
    Hint: One of the numbers must be even!

49. No.

50. 2:1.

51. 496.

52. $10^3 + 9^3$ and $1^3 + 12^3$.

53. *y*/12 cm.
    Hint: Solve for *h*: $1/3 \ \pi(x^2)y = \pi(2x)^2h$.

54. 20.
    Note that they are not all vertically or horizontally oriented.

55. 30.

56. 4/3.

$$1.010101 \ldots = 1 + \frac{1}{4} + \frac{1}{16} + \ldots = \frac{1}{1 - \frac{1}{4}}.$$

57. 3,999,960.
    The average number is 33333 and there are $5! = 120$ numbers.

58. 3032 or 3432.

59. 36.

60. 625.

61. 3.
    Then each pile would consist of 1/3 X cards and 2/3 O cards.

62. 0, 1, 4, 5, 6, and 9.

63. 3:π.

64. 32.

65. *N.*

66. 180°

67. An adjustment at midnight on 30 June will suffice until 1 October.

68. 14 sq. units.
    $A = (1 \cdot 2 + 2 \cdot 4 + 4 \cdot 8) - 1/2 \ (2 + 4 + 8) \ (4).$

69. 0.
    The presence of 2 and 5 will produce a multiple of 10.

70. 2 minutes.
    Note that in 1 hour, the hand moves through 30° of arc.

71. 329476.
    Note that perfect squares never end in 2, 3, 7, or 8.

72. (*a*) 16.

73. 29, 38, 47, 56, 65, 74, 83, and 92.
    Note that the digits sum to 11 in each case. You may want to try to explain why this must be the case.

74. 121°.

75. 0.
    Note that $(x - x)$ is a factor.

76. 2/3.

77. The $1.29 option.

78. 24.

79. 4, 5, 7, 9, 10 and 4, 6, 7, 8, 10 are possibile sets.

80. 4:3.

References

Contributors of the Problems

T he following list provides names of the contributors of the problems in the forms in which they appeared in the *Mathematics Teacher*. Contributions made by mathematics classes or groups of individuals are identified by the name of the first contributor, followed by *et al.*

After the names, the entries show the problems contributed, in the order in which they appear in this book. The sections are abbreviated as follows: Number Theory (NT), Coordinate Geometry (CG), Spatial Sense (SS), Logic (L), Algebra (A), Probability (PR), Geometry (G), Logs and Exponents (LE), Sequences and Patterns (SP), Counting (C), Word Problems (WP), Puzzles and Games (PZ), Facts (F), and Quickies (Q).

Adkins, Carmen. Q63.

Anderson, Lynn, et al. PZ1.

Arcidiacono, Mike, et al. NT14, SS2, SS14, SP4, SP5, C14.

Armistead, Joan. L13, SP14, C20, C23, WP2, WP18.

Barnett, Tracey, et al. WP8, PZ6, F9.

Barsby, John. NT17, CG5, CG07, LE2, C10, C16, WP10, Q9, Q25, Q38, Q75.

Beamer, James E., et al. G25, Q31.

Becker, Robert H. SS19, SS20.

Bell, Karen, et al. NT6, Q53.

Besancon, Susan L. G5, Q54.

Bettina, Joe. LE8.

Bhattacharya, Dipendra N., et al. WP3, WP4, PZ10.

Bridges, Linda. NT10, A18, LE12, C21.

Brosnan, Patricia A. PR9, G9, G34.

Brown, Richard G. G38.

Bunch, Mary Emma. Q48.

Bussey, Bruce. G13, G22, SP7, Q24.

Caniglia, Joanne. NT3, F4, F5, F6, F14, Q3, Q52.

Carter, Claudia, et al. SP2, Q10.

Carter, John. CG14.

Chapman, Kim. SS12, PR14, G16.

Chung, Samuel. C18.

Clarion University Problem Solving Class. L3, PZ6.

Crawford, Rudd, et al. PR13.

Cusick, Larry, and Tuska, Agnes. A21, G30, Q20.

Daniels, David S. CG9, G24, SP15, SP16, SP17, SP18, SP26, Q1, Q2, Q19.

Davies, Trefor. L14, PZ24.

DeLeon, Morris Jack. A2, LE15.

Ferucci, Beverly. G12.

Feser, Victor G. CG12, SS6, L1, A6, G15, G17, WP13, WP14, PZ9, Q58.

Findell, Carol. NT25, A15, A26, SP1, WP15, WP16, PZ22, Q9, Q18, Q32, Q47.

Foster, Alan G., and Theesfeld, Carole A. LE7, LE9, LE10, A16, A17, G10, G11, G14, G23, Q16, Q42, Q68, Q80.

Gorini, Catherine, et al. A33, F16.

Grant McLoughlin, John. NT8, NT15, NT16, NT23, NT26, NT29, NT30, SS21, L9, L11, CG1, CG8, A10, A11, A12, A14, A20, A22, A24, A25, A27, A28, PR2, PR4, PR5, PR6, PR7, PR15, PR16, PR19, LE3, LE14, LE18, LE23, SP3, SP11, SP12, SP13, SP24, C3, C5, C12, C19, WP11, PZ3, PZ16, PZ21, PZ23, PZ25, Q13, Q29, Q30, Q57.

Grant McLoughlin, Kathy. Q26.

Hamilton, Ilene. C2.

Harbison, Mark. Q14.

Heckman, David. Q44.

Henry, Boyd. NT1, NT19, PR8, PR22, C9.

Hsu, Li-Ling. LE6.

Hoggard, John. LE19.

Holdan, Gregory. SS5, SS18, SP10, WP5, PZ11.

Hull, Matthew. PZ7, PZ18.

Ionas, Elias. A7, A13, LE16, Q17.

Jamski, William D. NT13, Q67, Q70.

Kahle, Diane. NT4, Q61, Q74, Q77.

Kanold, Tim. A9.

Karytowski, William. CG2.

Kennedy, Joe. NT21, G31, G33, Q49.

Kenney, Bob Tex. G19, Q71.

Kenney, Margaret J., and Bezuszka, S. J. SS8, SS13, SS15, SS16, SS17, A1, PR3, G28, G29, C17, PZ19 Q21.

Kerr, Debra. CG4, SP6, SP25, PZ14, Q22, Q55.

Kinner, Bob. G32.

Kraus, William H. PZ13.

Lawson, Dene. A30, SP23, Q35.

Lewis, Allen. F1.

Little, Richard, et al. NT2, NT12, C7.

Lowman, Pauline, and Stokes, Joseph. NT20, CG6, L8, G26, C15.

Manning, Paul R. F7, Q37.

Markel, Kristopher. G7

McDonald, Jacqueline. L7.

Morgan, Denise. SS1, WP12.

Mowat, Elizabeth. CG11, SS10, C4, Q46, Q62, Q66, Q69.

Murphy, Micah. NT9.

Mutz, Grace. Q36.

Oliver, Scott. NT22, Q6.

Olson, Alton A., and Gordon, Lynn. SS9, G1, G2, G3, C13, Q79.

Otto, Albert. C11, PR11, NT5.

Phung, Scott. A34.

Pollock, Norman S. G4.

Raub-Hunt, Margaret, and Hunt, William. SS4, PR12, C2, PZ4, PZ15, F17, Q11.

Rauch, Kathie. LE20.

Rising, Gerry. A8, Q4, Q5.

Saul, Mark E. NT11, NT18, CG13, A5, A23, A31, A32, LE11, LE17, LE21, SP9, C22, Q7, Q8, Q23, Q33, Q41, Q50, Q56, Q59, Q76, Q78.

Schwieger, Ruben. G18, WP9, WP19, PZ8.

Scully, Barry. C6.

Shannon, Eileen. SS7, G39, F10, Q65, Q72.

Shultz, Harry S. NT31, L10, PR17, PR18, SP16, Q34, Q45.

Silverman, Neal. G35.

Simon, Harry. G8.

Slotta, Olive Ann, et al. F11.

Small, Marian. SP19, Q39, Q40, Q43, Q64.

Smith, Donald. F15, F18.

Smith, Tom. PR23.

Steuben, Michael A. L2, L4, L6, L12, PR24, WP1.

Swanson, Todd. A19, PR21, LE1, LE4, SP8, SP22, WP17, PZ17, Q60.

Swartz, Marybeth. Q28.

Taback, Stanley F., and Taback, Jennifer. G6, SP21, C1.

Tanyolu, Ebru. WP7, C8.

Thomas, Catherine, et al. SS3.

Thomson, Betty J. NT7, F8, Q27, Q51.

Tomhave, William K. NT24, NT27, NT28, SS11, L5, LE5, LE13, Q73.

Tuska, Agnes. PR20, PZ12.

Uva, Enrico. A3, A35, SP20.

Varnadore, James. CG3, A29, PR1, PR10, G20, G27, Q15.

Wagner, Doug. LE22.

Walton, Karen Doyle, and Walton, Zachary. CG10, A4, G21, G36.

Watts, John. NT2, NT12, C7.

Willis, Chris. F12, F13.

Wilson, Melvin R. PZ2, Q12.

Wong, Susan Knueven. F2, F3.

Zandy, Victor. PR13, PZ20.

Zuiker, Mark A. G37.

Calendar Dates of the Problems

A ll problems in this book were published in calendars in the *Mathematics Teacher* between October 1993 and May 1998. The calendar date of each problem appears in the following list, which is arranged by topic. Readers interested in particular problems may wish to check the original source for solutions and graphics. Occasionally, a problem has been reworded in the book, and many solutions published here are different from those that were originally printed with the calendars. The solutions to the Quickies in particular are likely to be more extensive in their original form. Readers can also refer to dates and check original sources for details pertaining to members of groups contributing problems.

Number Theory

NT01—7 November 1997
NT02—6 April 1995
NT03—9 March 1995
NT04—20 March 1994
NT05—1 December 1995
NT06—19 March 1996
NT07—1 May 1998
NT08—14 November 1996
NT09—2 February 1995
NT10—21 November 1995
NT11—14 February 1994
NT12—9 April 1995
NT13—12 May 1995
NT14—4 October 1996
NT15—8 November 1996
NT16—23 May 1997
NT17—23 January 1996
NT18—25 December 1993
NT19—24 November 1997
NT20—17 February 1995
NT21—11 October 1995
NT22—18 February 1996
NT23—15 January 1998
NT24—1 December 1996
NT25—27 November 1994
NT26—26 May 1998
NT27—21 December 1996
NT28—23 February 1997
NT29—20 January 1998
NT30—1 February 1998
NT31—3 March 1998

Coordinate Geometry

CG01—7 April 1996
CG02—6 November 1995
CG03—23 March 1994
CG04—14 November 1993
CG05—24 January 1996
CG06—22 January 1994
CG07—26 January 1996
CG08—26 October 1996
CG09—5 May 1996
CG10—7 March 1996
CG11—26 January 1997
CG12—21 December 1997
CG13—4 October 1993
CG14—13 March 1996

Spatial Sense

SS01—26 March 1995
SS02—1 October 1996
SS03—13 April 1994
SS04—8 December 1995
SS05—14 April 1995
SS06—19 December 1997
SS07—28 February 1997
SS08—10 May 1997
SS09—5 January 1997
SS10—20 January 1997
SS11—4 December 1996
SS12—19 May 1995
SS13—10 April 1997
SS14—17 November 1996
SS15—9 May 1997
SS16—19 April 1997
SS17—9 April 1997
SS18—28 May 1995
SS19—19 May 1997
SS20—20 May 1997
SS21—21 April 1997

Logic

L01—1 December 1997
L02—3 March 1997
L03—19 November 1993
L04—1 March 1997

L05—10 December 1996
L06—12 March 1998
L07—13 October 1994
L08—23 January 1994
L09—16 January 1998
L10—10 March 1998
L11—17 January 1998
L12—9 March 1998
L13—16 April 1994
L14—24 November 1994

Algebra

A01—14 April 1997
A02—20 October 1997
A03—15 October 1997
A04—2 February 1996
A05—20 February 1994
A06—23 December 1997
A07—22 February 1998
A08—24 January 1998
A09—14 March 1996
A10—19 April 1996
A11—5 February 1998
A12—21 November 1996
A13—3 January 1998
A14—2 February 1997
A15—10 May 1994
A16—25 October 1995
A17—23 October 1995
A18—11 November 1995
A19—3 April 1998
A20—4 October 1997
A21—26 December 1997
A22—4 February 1998
A23—28 January 1994
A24—31 October 1996
A25—28 October 1996
A26—5 May 1994
A27—18 May 1997
A28—3 February 1997
A29—24 February 1994
A30—21 November 1994
A31—29 November 1993
A32—30 October 1993
A33—12 February 1997
A34—10 May 1998
A35—15 November 1997

Probability

PR01—27 March 1994
PR02—18 April 1996
PR03—6 April 1997
PR04—25 December 1997
PR05—10 April 1996
PR06—1 January 1998
PR07—24 May 1997
PR08—9 November 1997

PR09—15 February 1997
PR10—24 March 1994
PR11—4 December 1995
PR12—21 May 21 1996
PR13—28 April 1995
PR14—8 April 1995
PR15—17 May 1997
PR16—9 November 1996
PR17—15 March 1998
PR18—1 March 1998
PR19—22 April 1996
PR20—10 February 1995
PR21—2 April 1998
PR22—12 November 1997
PR23—13 April 1995
PR24—4 March 1997

Geometry

G01—19 January 1997
G02—3 January 1997
G03—6 February 1997
G04—23 November 1994
G05—16 October 1996
G06—1 February 1997
G07—24 March 1995
G08—14 November 1994
G09—26 February 1997
G10—8 November 1995
G11—6 January 1996
G12—10 November 1994
G13—5 November 1996
G14—16 December 1995
G15—13 December 1997
G16—18 May 1995
G17—16 December 1997
G18—11 December 1994
G19—25 October 1997
G20—24 October 1994
G21—11 March 1996
G22—2 November 1996
G23—19 December 1995
G24—18 May 1996
G25—18 March 1997
G26—20 February 1995
G27—21 April 1994
G28—28 April 1997
G29—29 April 1997
G30—29 December 1997
G31—2 October 1995
G32—22 November 1997
G33—13 October 1995
G34—17 February 1997
G35—20 October 1993
G36—8 March 1996
G37—5 March 1994
G38—4 May 1997
G39—30 March 1997

Logs and Exponents

LE01—8 April 1998
LE02—18 January 1996
LE03—21 May 1997
LE04—13 April 1998
LE05—11 December 1996
LE06—26 February 1995
LE07—8 January 1996
LE08—8 February 1996
LE09—28 October 1995
LE10—20 October 1995
LE11—3 October 1993
LE12—22 November 1995
LE13—7 December 1996
LE14—22 May 1996
LE15—13 October 1997
LE16—6 January 1998
LE17—26 December 1993
LE18—20 April 1996
LE19—25 December 1994
LE20—11 February 1996
LE21—28 October 1993
LE22—4 February 1996
LE23—27 November 1996

Sequences and Patterns

SP01—25 October 1994
SP02—3 May 1997
SP03—11 April 1996
SP04—30 November 1996
SP05—3 October 1996
SP06—3 November 1993
SP07—4 November 1996
SP08—15 April 1998
SP09—24 November 1993
SP10—25 May 1995
SP11—20 April 1997
SP12—11 October 1996
SP13—4 February 1997
SP14—9 December 1995
SP15—12 May 1996
SP16—6 May 1996
SP17—15 May 1996
SP18—16 May 1996
SP19—23 March 1998
SP20—14 November 1997
SP21—24 February 1997
SP22—6 April 1998
SP23—1 November 1997
SP24—3 October 1997
SP25—11 December 1993
SP26—9 May 1996

Counting

C01—16 December 1996
C02—15 February 1996
C03—13 January 1998
C04—22 December 1996
C05—7 February 1998
C06—26 May 1997
C07—4 April 1995
C08—22 April 1995
C09—6 November 1997
C10—21 January 1996
C11—13 January 1996
C12—25 May 1996
C13—2 January 1997
C14—13 November 1996
C15—15 December 1993
C16—31 January 1996
C17—13 April 1997
C18—3 February 1995
C19—25 May 1997
C20—14 December 1993
C21—2 November 1995
C22—15 October 1993
C23—6 February 1995

Word Problems

WP01—5 March 1997
WP02—10 December 1995
WP03—26 October 1993
WP04—6 March 1996
WP05—16 April 1995
WP06—5 March 1996
WP07—21 April 1995
WP08—5 February 1994
WP09—7 January 1994
WP10—15 January 1996
WP11—20 November 1996
WP12—31 March 1995
WP13—10 January 1998
WP14—18 December 1997
WP15—18 May 1994
WP16—19 May 1994
WP17—9 April 1998
WP18—18 April 1994
WP19—13 December 1994

Puzzles and Games

PZ01—29 November 1997
PZ02—30 March 1996
PZ03—2 February 1998
PZ04—28 March 1996
PZ05—15 May 1998
PZ06—1 February 1994
PZ07—23 December 1994
PZ08—14 December 1994
PZ09—20 December 1997
PZ10—21 October 1993
PZ11—24 May 1995
PZ12—9 February 1995
PZ13—14 March 1997
PZ14—13 November 1993

PZ15—4 April 1996
PZ16—30 May 1997
PZ17—11 April 1998
PZ18—22 December 1994
PZ19—16 April 1997
PZ20—29 April 1995
PZ21—28 May 1998
PZ22—28 October 1994
PZ23—13 May 1998
PZ24—19 December 1994
PZ25—24 May 1998

Facts

F01—3 April 1996
F02—1 April 1996
F03—29 April 1996
F04—1 March 1995
F05—8 March 1995
F06—13 March 1995
F07—16 October 1995
F08—18 January 1994
F09—11 February 1994
F10—28 March 1997
F11—19 October 1993
F12—6 April 1996
F13—5 April 1996
F14—2 March 1995
F15—14 December 1995
F16—11 February 1997
F17—7 December 1995
F18—12 December 1995

Quickies

Q01—19 May 1996
Q02—14 May 1996
Q03—10 March 1995
Q04—27 January 1998
Q05—25 January 1998
Q06—17 February 1996
Q07—1 October 1993
Q08—31 October 1993
Q09—2 May 1994
Q10—8 May 1997
Q11—23 May 1996
Q12—26 March 1996
Q13—29 January 1998
Q14—29 October 1997
Q15—27 May 1994
Q16—2 January 1996
Q17—5 January 1998
Q18—16 May 1994
Q19—13 May 1996
Q20—28 December 1997
Q21—1 April 1997
Q22—4 November 1993
Q23—16 October 1993
Q24—3 November 1996

Q25—29 January 1996
Q26—23 November 1996
Q27—24 March 1997
Q28—6 December 1995
Q29—24 October 1996
Q30—22 January 1998
Q31—19 December 1996
Q32—4 May 1994
Q33—17 October 1993
Q34—16 March 1998
Q35—12 January 1994
Q36—12 April 1996
Q37—17 October 1995
Q38—20 January 1996
Q39—18 March 1998
Q40—20 March 1998
Q41—23 December 1993
Q42—29 October 1995
Q43—26 March 1998
Q44—26 December 1996
Q45—7 March 1998
Q46—24 December 1996
Q47—13 May 1994
Q48—24 April 1996
Q49—4 October 1995
Q50—2 October 1993
Q51—9 May 1998
Q52—7 March 1995
Q53—27 February 1996
Q54—8 October 1996
Q55—11 November 1993
Q56—26 January 1994
Q57—23 January 1998
Q58—30 December 1997
Q59—16 February 1994
Q60—16 April 1998
Q61—15 March 1994
Q62—28 December 1996
Q63—11 April 1995
Q64—24 March 1998
Q65—29 March 1997
Q66—31 January 1997
Q67—21 November 1997
Q68—22 October 1995
Q69—30 January 1997
Q70—15 January 1994
Q71—8 October 1997
Q72—9 February 1997
Q73—21 February 1997
Q74—13 March 1994
Q75—27 January 1996
Q76—28 December 1993
Q77—10 March 1994
Q78—27 December 1993
Q79—15 January 1997
Q80—21 October 1995

Bibliography

The following works have been cited as sources of problems that appear in this book. Citations can be found in parentheses in the corresponding solutions.

Billstein, Rick, Schlomo Libeskind, and Johnny Lott. *A Problem Solving Approach to Mathematics for Elementary School Teachers.* 5th ed. Reading, Mass.: Addison-Wesley, 1993.

Bolt, Brian. *Mathematical Funfair.* Cambridge: Cambridge University Press, 1989.

Brousseau, Alfred. *Mathematics Contest Problems.* Mountain View, Calif.: n.d.

Bruni, James V., and Helene J. Silverman. "Making Squares with a Pattern." *Arithmetic Teacher* 24 (1977): 265–72.

Devi, Shakuntala. *Puzzles to Puzzle You.* Delhi: Orient Paperback, 1990.

Dudeney, Henry Ernest. *Amusements in Mathematics.* New York: Dover Publications, 1958.

Dunn, Angela Fox. *Second Book of Mathematical Bafflers.* New York: Dover Publications, 1983.

Dunn, Angela, ed. *Mathematical Bafflers.* New York: Dover Publications, 1980.

Ecker, Michael W. *Getting Started in Problem Solving and Math Contests.* New York: Franklin Watts Publishing, 1987.

Eves, Howard. *An Introduction to the History of Mathematics.* New York: Holt, Rinehart, and Winston, 1964.

Friedland, Aaron J. *Puzzles in Math and Logic.* New York: Dover Publications, 1970.

Gardner, Martin. *The Numerology of Dr. Matrix.* New York: Simon and Schuster, 1967.

———. *The Unexpected Hanging and Other Mathematical Diversions.* Chicago: University of Chicago Press, 1991.

Gardner, Martin, ed. *The Mathematical Puzzles of Sam Loyd.* New York: Dover Publications, 1959.

Greenes, Carole. *Problem Resource Book.* Palo Alto, Calif.: Dale Seymour Publications, 1986.

Grossman, Jerrold W. *Discrete Mathematics.* New York: Macmillan Publishing Company, 1990.

Holt, Michael. *More Math Puzzles and Games.* New York: Walker and Company, 1978.

Joseph, George Gheverghese. *The Crest of the Peacock: Non-European Roots of Mathematics.* London: Penguin Books, 1992.

Lataille, Jane. "Problems 6, 10, 26; May 1998" [Letter to Reader Reflections]. *Mathematics Teacher* 92 (February 1999): 156.

Lindquist, Mary Montgomery. "Problem Solving with Five Easy Pieces." *Arithmetic Teacher* 25, no. 2 (1977): 6–10.

Osen, Lynn. *Women in Mathematics.* Cambridge, Mass.: MIT Press, 1974.

Pólya, George, and Jeremy Kilpatrick. *The Stanford Mathematics Problem Book: With Hints and Solutions.* New York: Teachers College Press, 1974.

Saint Mary's College Mathematics Contest Problems. Palo Alto, Calif.: Creative Publications, 1972.

Savin, Anatoly. "A Summer Festival of Puzzlers." *Quantum* 3 (July/August 1993): 32.

Summers, George J. *New Puzzles in Logical Deduction.* New York: Dover Publications, 1968.

Townsend, Charles Barry. *World's Toughest Puzzles.* New York: Sterling Publishing, 1990.

Trigg, Charles W. *Mathematical Quickies.* Mineola, N.Y.: Dover Publications, 1985.

Whimbey, Arthur, and Jack Lockhead. *Beyond Problem Solving and Comprehension: An Exploration of Quantitative Reasoning.* Hillsdale, N.J.: Lawrence Erlbaum Associates, 1984.

van der Waerden, Bartel Leendirt. *Geometry and Algebra in Ancient Civilations.* New York: Springer-Verlag, 1973.

Zabanch, Munirv. "Ideas on April 1992" [Letter to Reader Reflections]. *Mathematics Teacher* 86 (March 1993): 194–195.

Other Sources

Ainley, Stephen. *Mathematical Puzzles.* Englewood Cliffs, N.J.: Prentice Hall, 1977.

Barbeau, Edward. *After Math: Puzzles and Brainteasers*. Toronto: Wall and Emerson, 1995.

Barbeau, Edward J., Murray S. Klamkin, and William O. J. Moser, eds. *Five Hundred Mathematical Challenges*. Washington, D.C.: Mathematical Association of America, 1995.

Bates, Nathaniel B., and Sanderson M. Smith. *101 Puzzle Problems*. Concord, Mass.: Bates Publishing, 1995.

Booth, Peter, Bruce Shawyer, and John Grant McLoughlin. *Shaking Hands in Corner Brook and Other Math Problems—for Senior High School Students*. Waterloo, Ont.: Waterloo Mathematics Foundation, 1995.

Brooke, Maxey. *150 Puzzles in Crypt-Arithmetic*. New York: Dover Publications, 1963.

Clarke, Barry R. *Puzzles for Pleasure*. Cambridge: Cambridge University Press, 1995.

Flewelling, Gary. *Recreational Math Problems for High School Students*. Rev. ed. Bks. I and II. Guelph, Ont.: Grand Valley Mathematics Association, 1984.

Gardiner, Anthony. *Discovering Mathematics: The Art of Investigation*. Oxford: Oxford Science Publications, 1987.

Geretschlager, Robert, and Gottfried Perz. "Mathematics Competitions for the Under 15's in Austria." *Mathematics Competitions* 8, no. 2 (1993): 10–13.

Gilbert, George T., Mark J. Krusemeyer, and Loren C. Larson. *The Wohascum County Problem Book*. Washington, D.C.: Mathematical Association of America, 1993.

Holton, Derek. *Let's Solve Some Math Problems*. Waterloo, Ont.: Canadian Mathematics Competition, 1993.

Jackson, Bradley W., and Dorini Thoro. *Applied Combinatorics with Problem Solving*. Reading, Mass.: Addison-Wesley, 1990.

Jamski, William D., ed. *Mathematical Challenges for the Middle Grades from the* Arithmetic Teacher. Reston, Va.: National Council of Teachers of Mathematics, 1990.

Madachy, Joseph S. *Madachy's Mathematical Recreations*. New York: Dover Publications, 1979.

Perelman, Yakov. *Mathematics Can Be Fun*. Moscow: Mir Publishers, 1985.

Sgroi, Laura, and Rich Sgroi. *Mathematics for Elementary School Teachers: Problem Solving Investigation*. Newburgh, N.Y.: Newburgh Enlarged City School District, n.d.

Totten, Jim, ed. *Cariboo College High School Mathematics Contest: Problems, 1973–1992*. Kamloops, B.C.: Cariboo College, 1992.

Appendixes

Solutions to the
January 1995 Calendar

1. 47.12°.

    To convert a Celsius temperature C to a temperature T on the new scale, find the equation of a line through the freezing point, $(C, T) = (0, 19)$, and the boiling point, $(C, T) = (100, 95)$. $T = 0.76C + 19$ is the equation. Substituting $C = 37.0°$ gives $T = 47.12°$.

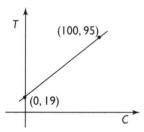

    An alternative solution would be to draw two thermometers to any convenient scale, numbering one up and the other down. Scales can differ, but the lines should be parallel. Next, connect 19 on the left thermometer to 0 on the right thermometer, and 95 on the left to 100 on the right. The pivot point is that where these two connectors cross. Next, draw a line from 37 degrees on the Celsius thermometer through the pivot. The point at which it crosses the new scale is the correct conversion. Why?

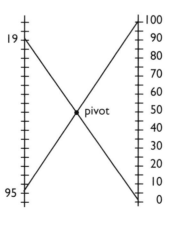

2. No.

    $40^2 + 19^2 + 5^2 + 3^2$ and $44^2 + 7^2 + 3^2 + 1^2$ are two of many possible answers.

3. 1995.

    The given table is a multiplication table. Since each row and column are represented just once by the three chosen numbers, their probuct always equals $1 \cdot 5 \cdot 7 \cdot 1 \cdot 3 \cdot 19 = 1995$.

	1	3	19
1	1	3	19
5	5	15	95
7	7	21	133

4. $-4, -3, -2, \ldots, 14$.

    Let the nineteen integers be $a - 9, a - 8, a - 7, \ldots, a - 1, a, a + 1, \ldots, a + 9$. Summing these integers, we get $19a = 95$. Therefore, $a = 5$ and the smallest integer is $a - 9 = -4$.

    Or average the nineteen consecutive numbers to their middle number. Thus, the middle number is 95/19, which is 5. So the smallest number is $5 - 9$, or -4, and the largest is $5 + 9$, or 14.

All problems for this very special calendar were provided by Richard G. Brown, Phillips Exeter Academy, Exeter, NH 03833.

5. The graph of the equation is a line with x-intercept 19 and with y-intercept 95. Since the slope is −5, the twenty points $(x, y) = (0, 95), (1, 90), (2, 85) \ldots, (19, 0)$ lie on this line.

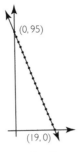

6. $x = 5, y = 3, z = 7$.

The volume of the solid is $19xyz = 1995$, so $xyz = 105 = 1 \cdot 3 \cdot 5 \cdot 7$. Since no edge is longer than 19, $4y \leq 19$, $3x \leq 19$, and $2z \leq 19$. Thus, $y \leq 4$, $x \leq 6$, and $z \leq 9$. The only possibility is $x = 5$, $y = 3$, and $z = 7$.

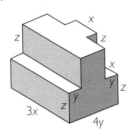

7. 6 and 8.

Multiply both sides of the equation by $F \cdot I \cdot V \cdot E$, getting

$$\frac{N \cdot I \cdot N \cdot E \cdot T \cdot E \cdot E \cdot N}{N \cdot I \cdot N \cdot E \cdot T \cdot Y} = O \cdot N \cdot E$$

Reduce the fraction, getting

$$\frac{E \cdot E \cdot N}{Y} = O \cdot N \cdot E$$

This equation simplifies to $E = O \cdot Y$. Since E, O, and Y are different, the only solutions are $6 = 2 \cdot 3$ or $3 \cdot 2$ and $8 = 2 \cdot 4$ or $4 \cdot 2$. Thus, E must be 6 or 8.

8. $\dfrac{361\sqrt{15}}{4}$, or approximately 350.

Label the triangle as shown in the figure. Since the perimeter is 95, $CA + CB = 76$ and, therefore, vertex C must lie on an ellipse with focal points A and B. This situation means that the altitude to side AB is maximum when $\triangle ABC$ is isosceles and $CA = CB = 38$. The altitude is then

$$\frac{19\sqrt{15}}{2},$$

and the area is $\dfrac{361\sqrt{15}}{4} \approx 350$.

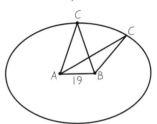

9. Any line through the center of a rectangle will separate it into equal areas. Therefore, the line through the centers of *both* rectangles will make $A + C = B + D$ and $C = D$. Hence, $A = B$. (Note that this solution is independent of the numbers 19 and 95.)

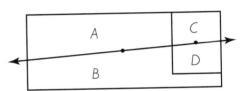

10. (b) On a rural road.

Note that 1995 cm/s is approximately 20 m/s or 72 km/h, or 45, MPH.

11. Row 22 and column 42.

The numbers in the first row are the triangular numbers $1, 3, 6, 10, 15, \ldots, n(n + 1)/2, \ldots$. These numbers are also the last numbers of the diagonals shown. To find the diagonal containing 1995, find the smallest triangular number greater than or equal to 1995. This number yields $n(n + 1)/2 \geq 1995$, whose smallest solution is $n = 63$. Thus, the 63rd entry in row 1 is the triangular number $63 \cdot 64/2 = 2016$. Since $2016 − 1995 = 21$, we must count twenty-one entries backward along a diagonal moving down and left. This movement takes us from the position (row, column) $= (1, 63)$ to $(1 + 21, 63 − 21) = (22, 42)$.

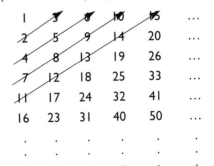

12. 7/12.

$$\frac{\text{area of one scalene } \triangle}{\text{area of big equalateral } \triangle}$$

$$= \frac{\dfrac{1}{2} \cdot 19 \cdot 95 \cdot \sin 60}{\dfrac{1}{2} \cdot 114 \cdot 114 \cdot \sin 60}$$

$$= \frac{5}{36}$$

Therefore, the total area of the three scalene triangles is $3 \times 5/36 = 5/12$ of the area of the big equilateral triangle. Thus, the small equilateral triangle has 7/12 of the area of the big one.

13. 2006.

A normal year is 52 weeks + 1 day (7 · 52 + 1 = 365 days). So the day of the week on which any particular January date falls advances one day each year and two days after a leap year (1996, 2000, 2004, ...). Remember that centennial years are leap years only when divisible by 400. Therefore, 2000 *is* a leap year along with 1996 and 2004. Thus, the progression for 13 January is the following:

1995 Friday	2001 Saturday
1996 Saturday	2002 Sunday
1997 Monday	2003 Monday
1998 Tuesday	2004 Tuesday
1999 Wednesday	2005 Thursday
2000 Thursday	2006 Friday

14. $19 + 9\sqrt{5}$.

Let us call the three-step left-up-diagonal movement a subpath. Its length is $2 + \sqrt{5}$, and it moves two units closer to the y-axis. Nine of these subpaths will have length $18 + 9\sqrt{5}$, ending just one unit from the y-axis. The last move left to the y-axis makes the total distance $19 + 9\sqrt{5}$.

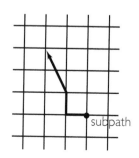

subpath

15. 18 to 1.

If t = the number of teachers and s = the number of students, then

$$0.19t + 0.95s = 0.91(t + s),$$

$$0.04s = 0.72t,$$

$$\frac{s}{t} = \frac{18}{1}.$$

16. (a) and (d).

(a) simplifies to $\log_5(1/5) = -1$, (b) and (c) are irrational, and (d) simplifies to $\log_5 |-5| = 1$.

17. 9.5 is both the maximum and the minimum possible length of DE.

Since $\triangle BEC$ is a right triangle with fixed hypotenuse BC and a right angle at E, E must lie on a circle with diameter BC. The center of the circle is D, and the radius is $DE = BC/2 = 19/2 = 9.5$.

Alternatively, since BE is an altitude, $\triangle BEC$ is a right triangle with hypotenuse BC. Since AD is the altitude to the base of an isosceles triangle, D is the midpoint of BC. Since the midpoint of the hypotenuse of a right triangle is equidistant from the vertices, $DE = BC/2 = 19/2 = 9.5$.

18. $19.95.

If p is the usual price, then $0.90p \times 5$ is the sale price for five books. The 3.8 percent tax brings the price to $5(1.038)(0.90p)$. Adding $2.00 to wrap the four gifts makes the final total $5(1.038)(0.90p) + 2 = 95.19$. Solving this equation gives $p = \$19.95$.

19. $(19, 95)$.

The east-west component of the journey is given by this infinite series:

$$95 - 95 \cdot \frac{1}{4} + 95 \cdot \frac{1}{16} - 95 \cdot \frac{1}{64} + \cdots$$

$$= 95\left(1 - \frac{1}{4} + \frac{1}{16} - \frac{1}{64} + \cdots\right) = \frac{95}{1 - \left(-\dfrac{1}{4}\right)} = 76.$$

The north-south component of the journey is given by this infinite series:

$$190 - 190 \cdot \frac{1}{4} + 190 \cdot \frac{1}{16} - 190 \cdot \frac{1}{64} + \cdots$$

$$= 190\left(1 - \frac{1}{4} + \frac{1}{16} - \frac{1}{64} + \cdots\right) = \frac{190}{1 - \left(-\dfrac{1}{4}\right)} = 152.$$

Thus, the ultimate destination point is

Starting point + (76, 152)
= (−57, −57) + (76, 152) = (19, 95).

20. After repeating step 2 a few times, the number 1995 appears over and over. To see why, suppose the calculator shows x over and over. Then step 2 indicates that

$$\sqrt{x + 3978030} = x.$$

Solving gives

$$x^2 - x - 3978030 = 0,$$
$$(x - 1995)(x + 1994) = 0.$$

Since $x > 0$, $x = 1995$.

21. $\dfrac{10^{2n} + 1}{5}$.

If $n = 1$, 1919/95 = 20.2; if $n = 2$, 19191919/9595 = 2000.2; if $n = 3$, 191919191919/959595 = 200000.2; and if $n = 4$, 1919191919191919/95959595 = 20000000.2. The answer here is always a 2 followed by $(2n − 1)$ zeros and 0.2, or $2 \times 10^{2n-1} + 0.2$, which is equivalent to $(10^{2n} + 1)/5$.

22. 19 and 95.
Reflect B across line CD to B' so that $PB = PB'$. Since $AP + PB$ is minimum, then so is $AP + PB'$, which means that A, P, and B' are collinear, making $\triangle ACP \sim \triangle B'DP$, with a scale factor 13/65 = 1/5. Thus, $CP/PD = x/(114 − x) = 1/5$, so $x = 19$ and $114 − x = 95$.

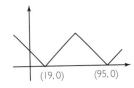

23. 19 and 95.

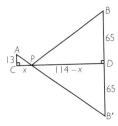

Alternatively,

$$y = \begin{cases} \big||x - 57| - 38\big|, & x \geq 57 \\ \big||57 - x| - 38\big|, & x < 57; \end{cases}$$

$$\big||x - 57| - 38\big| = 0,$$
$$x - 95 = 0,$$
$$x = 95,$$

or

$$\big||57 - x| - 38\big| = 0,$$
$$19 - x = 0$$
$$x = 19.$$

24. 1995.
$r^2 + rs + s^2$
= $(r + s)^2 − rs$
= (sum of roots)^2 − (product of roots)
= $50^2 − 505$
= 1995

25. The triangles are congruent, each having altitude 1995.
Q is the *center* of the first triangle, and 665 is 1/3 of the altitude, so the altitude is 1995. If a point P inside an equilateral triangle is *not* the center of the triangle, then let x, y, and z be the distances from P to the sides of the triangle. Also let b and h represent the base and altitude of the triangle. Then the area of $\triangle ABC$ is $(1/2)bh$, but it is also the sum of the areas of $\triangle ABP$, $\triangle BCP$, and $\triangle CAP$. Thus,

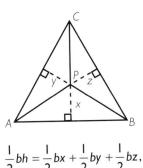

$$\frac{1}{2}bh = \frac{1}{2}bx + \frac{1}{2}by + \frac{1}{2}bz,$$
$$h = x + y + z$$
$$= 664 + 665 + 666$$
$$= 1995.$$

26. Sequences A and B can be described by the functions $a(n) = 19 + 76n$, $n = 0, 1, 2, \ldots$, and $b(m) = 20 + 25m$, $m = 0, 1, 2, \ldots$. To find a number in both sequences, we must find values of n and m such that

$$a(n) = b(m),$$
$$19 + 76n = 20 + 25m,$$
$$m = \frac{76n - 1}{25}$$
$$= \frac{75n + (n-1)}{25}.$$

The last equation indicates that $(n-1)$ must be divisible by 25, so $n = 1, 26, 51, \ldots$. Substituting $n = 1$ and $n = 26$ into $a(n) = 19 + 76n$ gives 95 and 1995, the first two numbers of sequence A that are also in sequence B.

27. Divide the perimeter of the cake, 38 inches, into 19 parts, each with length 2 inches. Five such pieces are shown. Note the right-hand corner piece. If the height of the cake is h, then the frosting on the side of each piece is $2h$ and the frosting on the top is $1/2(2)4.75$. The volume of each piece is $4.75h$.

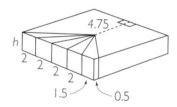

28. (1) 19; (2) 95.
The ratio of the volumes of two similar figures is the cube of their similarity ratio. Therefore,
$$\frac{\text{vol } A}{\text{vol}(A + B)} = \left(\frac{1}{2}\right)^3 = \frac{1}{8}.$$

Also,
$$\frac{\text{vol}(A + B)}{\text{vol}(A + B + C)} = \left(\frac{2}{3}\right)^3 = \frac{8}{27},$$

so
$$\frac{\text{vol}(A + B)}{\text{vol } C} = \frac{8}{19}.$$

Multiplying the first and third of these ratios gives
$$\frac{\text{vol } A}{\text{vol } C} = \frac{1}{8} \cdot \frac{8}{19} = \frac{1}{19},$$

so
$$\frac{\text{vol } C}{\text{vol } A} = 19,$$

and
$$\text{vol } C = 19 \times \text{vol } A$$
$$= 19 \times 5 = 95.$$

29. 997.
If $x^2 - y^2 = 1995$, then $(x + y)(x - y) = 1995$. This equation yields the pair of equations $x + y = A$ and $x - y = B$, where A and B are integers with a product of 1995. Solving these equations for y gives $y = (A - B)/2$, which is maximum when A is as large as possible and B is as small as possible. Therefore, $A = 1995$ and $B = 1$, so $y = (A - B)/2 = (1995 - 1)/2 = 997$.

30. $1995 = 7 + 8 + 9 + \ldots + 63$, the sum of fifty-seven consecutive positive integers.
If the first of the n consecutive integers is a, then
$$a + (a + 1) + (a + 2) + \ldots + (a + n - 1) = 1995,$$
$$\frac{n}{2}(2a + n - 1) = 1995,$$
$$n(2a + n - 1) = 3990$$
$$= 1 \cdot 2 \cdot 3 \cdot 5 \cdot 7 \cdot 19.$$

This last equation implies that
(1) n is a factor of 3990,

and

(2) $2a + n - 1 = 3990/n$

so that
$$a = (1995/n) - (n - 1)/2.$$
Using (1) and (2), we must maximize n at the same time that a is positive. Using trial and error, we can test $n = 7 \cdot 19 = 133$. But we find that this makes a less than 0, since from (2), $a = 1995/133 - 132/2 = -51$. Further trial and error yields the maximum $n = 3 \cdot 19 = 57$ and $a = 7$. The resulting sum of fifty-seven consecutive positive integers is $7 + 8 + 9 + \ldots + 63 = 1995$.

31. 1/1995.
A total of $2 \times 2992 + 1 = 5985$ numbers can be picked. To be divisible by 1995, a number must be divisible by 3, 5, 7, and 19, the prime factors of 1995. The number 29! satisfies this requirement. The only other numbers to do so are $29! - 1995$ and $29! + 1995$. Therefore, the probability that the number picked is divisible by 1995 is 3/5985 = 1/1995.

Appendix B
Recommended Reading

The shelves of a puzzler's personal library are usually rich with examples of wonderful problems and ideas for mathematical problem solving. Resources from contests, such as those available through the Canadian Mathematics Competition at the University of Waterloo, offer abundant ideas. The *Journal of Recreational Mathematics* is another source of mathematical entertainment. Books by Martin Gardner, David Wells, H. E. Dudeney, John Conway, Richard Guy, Ross Honsberger, Sam Loyd, Doris Schattschneider, Raymond Smullyan, Theoni Pappas, Ivars Peterson, and Ian Stewart are among the many that provide interesting bases from which to explore mathematical ideas or tackle specific problems.

The accompanying list of resources includes titles that are less likely to be familiar to many readers. Mathematical problem solvers are influenced by different individuals along the way. My own problem solving has been influenced greatly by Ed Barbeau. My understanding of the "educational potential" of problems was enriched and challenged through graduate courses and conversations with Steve Brown. In addition, my appreciation of the role of problems in mathematics learning has been enhanced and developed through numerous communications with Peter Taylor. The recognition of the roles of these individuals is reflected in my inclusion of some of their works in the list of suggested resources. Each problem solver would surely put together a different list. The authors whom I have just mentioned and the others whom I have included in the following bibliography offer many other suggestions.

Barbeau, Ed. *Power Play*. Washington, D.C.: Mathematical Association of America, 1997.

Barbeau, Ed. *After Math: Puzzles and Brainteasers*. Toronto, Ont.: Wall and Emerson, 1995.

Bolt, Brian. *Mathematical Funfair*. Cambridge: Cambridge University Press, 1989.

Booth, Peter, Bruce Shawyer, and John Grant McLoughlin. *Shaking Hands in Corner Brook and Other Math Problems—for Senior High School Students*. Waterloo, Ont.: Waterloo Mathematics Foundation, 1995.

Brown, Stephen I., and Marion I. Walter. *The Art of Problem Posing*. 2nd ed. Hillsdale, New Jersey: Lawrence Erlbaum Associates, 1990.

Cornelius, Michael, and Alan Parr. *What's Your Game? A Resource for Mathematical Activities*. Cambridge: Cambridge University Press, 1991.

Dunn, Angela, ed. *Mathematical Bafflers*. New York: Dover Publications, 1980.

Gardner, Martin. *The Magic Numbers of Dr. Matrix*. Buffalo, N.Y.: Prometheus Books, 1985.

Gardner, Martin. *The Mathematical Puzzles of Sam Loyd*. New York: Dover Publications, 1959.

Guy, Richard K., and Robert E. Woodrow, eds. *The Lighter Side of Mathematics: Proceedings of the Eugene Strens Memorial Conference on Recreational Mathematics and its History*. Washington, D.C.: Mathematical Association of America, 1994.

Hill, Thomas J. *Mathematical Challenges II—plus Six*. Reston, Virginia: National Council of Teachers of Mathematics, 1974.

Holton, Derek. *Let's Solve Some Mathematics Problems*. Waterloo, Ont.: Waterloo Mathematics Foundation, 1993.

Honsberger, Ross. *Mathematical Gems III*. Washington, D.C.: Mathematical Association of America, 1985.

Kraitchik, Maurice. *Mathematical Recreations*. New York: Dover Publications, 1953.

Mason, John, Leone Burton, and Kaye Stacey. *Thinking Mathematically*. London: Addison-Wesley, 1990.

Perelman, Yakov. *Mathematics Can Be Fun*. Moscow: Mir Publishers, 1985.

Posamentier, Alfred S., and Jay Stepelman. *Teaching Secondary Mathematics: Techniques and Enrichment Units*. 5th ed. Upper Saddle River, N.J.: Prentice Hall, 1999.

Schuh, Fred. *The Master Book of the Mathematical Recreations*. New York: Dover Publications, 1968.

Taylor, Peter D. "A Senior High School Math Textbook: In Progress." Kingston, Ont.: Queen's University 1998.

Totten, Jim, ed. *Cariboo College High School Mathematics Contest: Problems 1973–1992*. Kamloops, B.C.: Cariboo College, 1992.

Trigg, Charles W. *Mathematical Quickies*. Mineola, N.Y.: Dover Publications, 1985.

Vakil, Ravi. *A Mathematical Mosaic: Patterns and Solving*. Burlington, Ont.: Brendan Kelly Publishing, 1997.

Wells, David. *The Penguin Dictionary of Curious and Interesting Numbers*. London: Penguin, 1986.

Appendix C
Calendar Reviewers

Three individuals review every calendar problem for the *Mathematics Teacher*. The editor receives the comments of the reviewers along with copies of the problems that they have examined. The following list includes the names of all individuals who served as reviewers for any problem that appears in this collection.

Beers, George
Bhattacharya, Dipendra N.
Bode, John N.
Bollenbacher, Duane
Brown, Richard G.
Bruckner, Barry
Citynes, Michael
Collins, Jean
Commander, Emory S.
Contino, Michael A.
Cook, Marcy
Cooper, Judith M.
DePew, Nancy
Dusterhoff, Marilane
Gardenhire, Gene
Gardenhire, Susan
Giambrone, Tom M.
Grant McLoughlin, John
Henry, Boyd
Hoehn, Larry
Hynes, Michael C.
Jaffee, Mark
Jamski, William D.
Jones, Patricia L.
Kennedy, Bill
Kennedy, Joe
Kenney, Margaret J.
Kissack, Glenn
Kyser, Carrie
Leonard, Bill
Levy, Norton
Lin, Shinemin
Lindenberg, Mary
Little, Richard A.
Llewellyn, David W.

Marquis, June
Morrison, Mary Dell
Musick, Oleta
Obermann, Carol L.
Parker, Melanie
Petranek, Jerry J.
Polhamus, Edward C., Jr.
Riggle, Thomas
Savir, Etan
Scheetz, Peter
Sgarlotti, Richard
Sherrill, James
Skow, Donald P.
Smith, John M.
Snipes, Jeanette
St. Martin, Steven
Stevens, Jill
Stull, R. Scott
Tayeh, Carla
Thompson, E. Otis
Varnadore, Jim
Veldkamp, Arnold H.
Vennebush, Patrick
Vinson, Richard
Walton, Karen Doyle
Ware, James G.
Weier, C. Timothy
Weise, Jerry W.
Welch, Gary G.
Wexler, Steve
Woods, James E.
Wright, Stuart
Wyllie, Richard
Zerger, Monte J.
Zirkel, Gene